Advanced Structured Materials

Volume 160

Common engineering materials reach in many applications their limits and new developments are required to fulfil increasing demands on engineering materials. The performance of materials can be increased by combining different materials to achieve better properties than a single constituent or by shaping the material or constituents in a specific structure. The interaction between material and structure may arise on different length scales, such as micro-, meso- or macroscale, and offers possible applications in quite diverse fields.

This book series addresses the fundamental relationship between materials and their structure on the overall properties (e.g. mechanical, thermal, chemical or magnetic etc.) and applications.

The topics of *Advanced Structured Materials* include but are not limited to

- classical fibre-reinforced composites (e.g. glass, carbon or Aramid reinforced plastics)
- metal matrix composites (MMCs)
- micro porous composites
- micro channel materials
- multilayered materials
- cellular materials (e.g., metallic or polymer foams, sponges, hollow sphere structures)
- porous materials
- truss structures
- nanocomposite materials
- biomaterials
- nanoporous metals
- concrete
- coated materials
- smart materials

Advanced Structured Materials is indexed in Google Scholar and Scopus.

More information about this series at https://link.springer.com/bookseries/8611

J. M. P. Q. Delgado · A. G. Barbosa de Lima ·
Laura H. Carvalho

Moisture Transport in Polymer Composite Materials

Computational Modelling and Experiments

Springer

J. M. P. Q. Delgado
CONSTRUCT-LFC
Department of Civil Engineering
Faculty of Engineering
University of Porto
Porto, Portugal

A. G. Barbosa de Lima
Department of Mechanical Engineering
Federal University of Campina Grande
Campina Grande, Paraíba, Brazil

Laura H. Carvalho
Department of Mechanical Engineering
Federal University of Campina Grande
Campina Grande, Paraíba, Brazil

ISSN 1869-8433 ISSN 1869-8441 (electronic)
Advanced Structured Materials
ISBN 978-3-030-77825-5 ISBN 978-3-030-77826-2 (eBook)
https://doi.org/10.1007/978-3-030-77826-2

This Springer imprint is published by the registered company Springer Nature Switzerland AG
The registered company address is: Gewerbestrasse 11, 6330 Cham, Switzerland

Preface

Professionals from chemical, materials, mechanical and manufacturing engineering, as well as from other scientific areas are responsible for the great advance in new materials, particularly in the field of polymer composites.

Nowadays, the development of new and advanced fiber-reinforced polymer composites has allowed the manufacture of innovative products with high quality from the technological and mechanical points of view. These products are manufactured by the automotive, marine, aerospace, and sports products industries. A significant amount of research on polymer composites deals with thermoset vegetable fiber-reinforced systems manufactured by hand lay-up. In fact, these materials have gained preference because they offer some advantages such as easy processing and manufacturing in products of different shapes, quality and good mechanical properties of the manufactured products and high potential for diverse industrial and technological applications, especially those associated with socio-economic and environmental benefits. However, raw materials from renewable sources, such as vegetable fibers, present some disadvantages when used as reinforcement in polymers, due to their strong hydrophilic nature and weak interfacial adhesion with most polymer matrices, which are responsible for reducing the mechanical properties of polymer composites.

The strong tendency of vegetable fiber-reinforced polymer composites to absorb water plays an important role to address studies related to the water absorption process in these materials. Several specialized journals report experimental data on this topic and most of them analyze the results by simple modeling. Unfortunately, this issue is hardly discussed in textbooks, which leads to a gap and the need for new academic books aimed at this issue, especially on topics related to non-traditional vegetable fibers and advanced mathematical modeling. Therefore, this book is intended to provide valuable information about the water absorption process in vegetable fiber-reinforced thermoset polymer composites with emphasis in experiments, modeling and simulation. In this document, emerging topics related to theory, engineering applications, advanced mathematical modeling (anomalous diffusion) and experiments on the water absorption process in polymer composites are presented and discussed.

For many years, we have the enormous pleasure of working with various talented researchers including our current and former students (undergraduate, graduate and post-graduate levels), in the themes "fiber-reinforced polymer composite and their related topics". This book documents our research progress in this field. The book has six chapters about water absorption process in composite materials, particularly of vegetable fiber-reinforced polymer composites.

Chapter 1 provides an introduction on fiber-reinforced polymer composites and the motivation to study the water absorption process in these materials.

Chapter 2 deals with the water absorption process in vegetable fiber-reinforced polymer composite materials. The main issues, including foundations, advantages and problems associated with the use of vegetable fiber-polymer composites, especially those provoked by weak adhesion and exposure to hot and humid environments are described.

Chapter 3 presents the experimental techniques developed by our research group related to water absorption in vegetable fiber-reinforced polymer composites. Herein, important information related to fiber morphology, composite manufacturing, moisture absorption, and mechanical characterization tests in polymer composites reinforced by caroá, macambira and sisal fibers are detailed.

Chapter 4 presents a comprehensive and rigorous analysis about pure diffusion through fibrous porous media (Fick's second law), with particular emphasis to the process of water absorption in vegetable fiber-reinforced polymer composites. In this chapter different approaches (analytical, numerical and by CFD) and their limitations for the correctly describing the water absorption process in vegetable fiber-reinforced polymer composites are given. Herein, a homogeneous and three-dimensional mathematical modeling that includes different effects of some process parameters (sample thickness and water bath temperature) in the moisture migration behavior inside the material is presented, with especial emphasis to polyester composites reinforced by caroá, macambira and sisal fibers.

Chapter 5 presents a rigorous theoretical approach to the anomalous diffusion through porous fibrous media (Langmuir-type model), its advantages and limitations. An advanced macroscopic mathematical modeling that considers the existence of water molecules in the free and entrapped states inside the material, during the process of water absorption is proposed. In this chapter, we perform different approaches (analytical and numerical) for the correct prediction of the process of water absorption in vegetable fiber-reinforced polymers. Here, we report applications on caroá fiber-reinforced polyester composites and other arbitrary cases, including the effect of water layer thickness at the surface of the composite on the overall water pick-up.

In Chap. 6, we present conclusions on the different chapter contents and the main results present in them. The idea is to help professionals, engineers, industrials and academics involved in this advanced and interdisciplinary field.

We would like to thank Prof. Dr. Wilma Sales Cavalcanti, Múcio Marcos Silva Nóbrega, Valério Carlos de Almeida Cruz, Danielton Gomes dos Santos, João Baptista da Costa Agra Melo, Wanessa Raphaella Gomes dos Santos, Ana Flávia Camara Bezerra, Carlota Joaquina e Silva and Rafaela Quinto da Costa Melo for providing excellent data on the subject treated here and contributions in this book.

Porto, Portugal — J. M. P. Q. Delgado
Campina Grande, Brazil — A. G. Barbosa de Lima
Campina Grande, Brazil — Laura H. Carvalho

Acknowledgements

We thank our Publisher and Editors for making valuable suggestions to the contents of the book, and especially to Prof. Dr.-Ing. Andreas Öchsner for believing in this project. Finally, we express our gratitude for each one of the cited authors and researchers, and to the Portuguese and Brazilian (CNPq, CAPES, FAPESQ-PB and FINEP) Research Agencies for the financial support.

Contents

1 Introduction 1
1.1 Motivation 1
1.2 Composite Materials: Fundamentals 2
References 6

2 Fundamentals 9
2.1 Natural Fiber-Reinforced Polymer Composites 9
2.2 Water Absorption in Natural Fiber-Reinforced Polymer Composites 12
References 14

3 Experimental Analysis 17
3.1 Background 17
3.2 Experimental Setup 18
3.3 Experimental Results Analysis 22
3.3.1 Morphology Characterization of the Fibers 22
3.3.2 Water Absorption Kinetics 23
3.3.3 Mechanical Properties 25
References 35

4 Fick's Model Analysis 39
4.1 Transport Phenomena in Porous Media: Foundations 39
4.2 Moisture Absorption by Fick's Model 40
4.2.1 The Physical Problem 40
4.2.2 The General Mass Diffusion Equation 40
4.2.3 The Mass Diffusion Equation: 3D Approach in Cartesian Coordinates 42
4.2.4 Solution Techniques: Three-Dimensional Approach 44
4.2.5 Fick's Model Application 49
References 65

5 Langmuir-Type Model Analysis ... 69
5.1 Fundamentals ... 69
5.2 Moisture Absorption by Langmuir-Type Model ... 71
5.2.1 The Physical Problem ... 71
5.2.2 The General Mass Diffusion Equation ... 71
5.2.3 The Mass Diffusion Equation: 3D Approach in Cartesian Coordinates ... 72
5.2.4 The Water Absorption Process: 1D Approach in Cartesian Coordinates ... 74
5.2.5 The Water Absorption Process: 3D Approach ... 90
References ... 97

6 Conclusions ... 99

Chapter 1
Introduction

This chapter provides several information about polymer composites with emphasis on vegetable fiber-reinforced polymer composite materials, as well the motivation to study the water absorption process in these materials. Herein, we present and discuss different topics such as definition, classification, constituents, technological characteristics, manufacturing techniques, performance and applications of polymer composites.

1.1 Motivation

Both academy and industry have given special attention to understand and predict the properties of composite materials as these materials have a wide range of applications in different sectors such as: boats, ships, automotive and aircraft components, submersibles, offshore structures and prostheses. Fiber reinforced polymer composite materials have been preferred to designing structural materials when compared to conventional materials (metals and ceramics), especialy that originated from vegetable fiber as reinforcement. The reasons for replacing syntetic fibers in polymer composites are the good set of mechanical properties (strength and stiffness) displayed by vegetable fibers, their availability, low cost, low density as well as other characteristics.

The mechanical properties of composite materials strongly depend on good adhesion between fiber and polymer matrix, which in turn, strongly depend on the manufacturing techniques employed to make them and on the operational and environmental conditions during use. One of these conditions is moisture absorption, especially at high temperatures.

Despite of its importance, few books offer detailed information about the water absorption process in vegetable-fiber reinforced polymer composites including experiments, mathematical models and their analytical and numerical solutions.

J. M. P. Q. Delgado et al., *Moisture Transport in Polymer Composite Materials*, Advanced Structured Materials 160, https://doi.org/10.1007/978-3-030-77826-2_1

Mathematical modeling and numerical computation are processes to obtain solutions for physical problems of academic and/or industrial interest and these include differential equations of mass and energy conservations. In these areas, books dedicated to this theme are rather scarce and limited to providing an understanding on this issue only under rectilinear infiltration using Fick´s second law of diffusion. However, is some situations, polymer composites display an anomalous behavior during the water absorprion process. In these cases, Fickian diffusion fails, particularly in the final stages of the process. Information addressed to the physical problems associated with the water uptake in polymer composite materials involve partial to non-Fickian moisture absorption. This information is almost null when dealing with polymer composites reinforced by vegetable fibers in 3D-geometry.

This comprehensive book aims to respond to the large growing interest associated with Fickian and non-Fickian moisture absorption processes in polymer composites, especially that involving advanced mathematical treatment of the governing equations, experiments and numerical computation. It is the first book entirely dedicated to this important subject.

The key challenge of this book is to document different information related to fiber-reinforced polymer composites ranging from basic material and manufacturing to advanced mathematical modeling of water migration in fibrous media, their effect, and applications, in a unique volume. This book contains information about Fickian and non-Fickian diffusion of water inside vegetable-fiber reinforced polymer composites, a subject not yet treated and discussed together.

The goal is to provide an in-depth analysis of the key issues including rigorous and coupled engineering models, theoretical and experimental results, and general fundamentals about water absorption in polymer composites reinforced by vegetable fiber. Thus, this book will assist professionals, curious readers, undergraduate and graduate students, as well as applied mathematicians, engineers and scientists to better understand advanced topics associated with fiber-reinforced polymer composites especially those related to the moisture absorption process.

Further, the authors sincerely believe that this book becomes an excellent reference source for professionals already cited and a start point for encouraging different people to study both polymer composite materials and the effect of water absorption in these materials.

1.2 Composite Materials: Fundamentals

Composites are multiphase materials consisting of one or more discontinuous phases (reinforcing filler), embedded in a continuous phase (matrix) [1–7]. Based in this definition, it is a new material obtained by combination of two or more materials insoluble in each other, with specific properties not found in either material alone.

Composites can be classified according to the type of matrix, the type, geometry and shape of the reinforcement used. In general, inorganic materials are used as reinforcement in organic matrices. Thus, composite materials can be classified as fiber-reinforced composites (fibrous composites) and particle-reinforced composites (particulate composites). Sometimes metallic wires and ribbons are used as reinforcement in composites.

Composite materials can also be classified according to the chemical and physical nature of the matrix which can be: ceramic, metallic or polymeric. Ceramic materials are inorganic in nature, have high resistance to heat and are fragile. Metallic materials, in general, present high ductility and excellent thermal and electrical conductivities. The great limitation in the use of metallic materials as reinforcement in composites is their high density and manufacturing costs. On the other hand, polymeric materials stand out for their low density, easy conformation and high electrical resistivity [8, 9].

Composites properties are strongly dependent on the properties of their constituents (matrix and reinforcement). In general, reinforcing materials are harder and stronger than the matrix. Thus, these materials are added to the matrix in order to increase some of its properties so as to achieve the best properties for a given application.

The shape, size, distribution, content and filler (reinforcement) orientation as well as the matrix/filler interfacial bonding strongly affect the initial and long term properties (hydro, thermal, electrical and mechanical) of the composites whose performance can also be affected by environmental exposure resulting in volumetric variations (swelling). Thus, it is very important to know these materials (matrix and reinforcement) in detail.

Depending on the nature and geometry of the reinforcement, the following classification applies:

(a) Nature: These materials must be hard to promote increase in hardness and abrasion resistance; rupture resistant to provide high tensile, flexural and shear strength; rigid to increase elastic modulus; flexible to increase impact resistance, and heat resistant to increase thermal stability of the composite.
(b) Geometric characteristics: These materials can be particulate, fibrous, and laminates.

Further, fiber reirforcements (long or short), globular particles and platelets can be incorporated into ductile or brittle matrices at random or in oriented laminates, generating composites with different structures and properties [8–14].

The reinforcing material can be a synthetic fiber (glass, carbon, aramid, etc.) or a natural fiber (sisal, ramie, jute, cotton, kenaf, pineapple, etc.) or particles (clay, mica, tungsten carbide, titanium carbide, etc.). Mineral fillers as reinforcements, often can be incorporated into matrix in order to reduce costs, especially when associated to ultra high strength fibers. Figure 1.1 illustrates a scheme for the classification and types of composite materials.

The matrix is responsible for the external appearance of the composite and the protection of the reinforcement against chemical and physical attacks. The fillers may be well dispersed of agglomerated in the composite matrix. However, when subjected to stresses, the matrix must deform sufficiently in order to adequately distribute and transfer stresses to the reinforcing material. The proper choice of matrix and reinforcement to be used in a structural application is defined based on the strain that

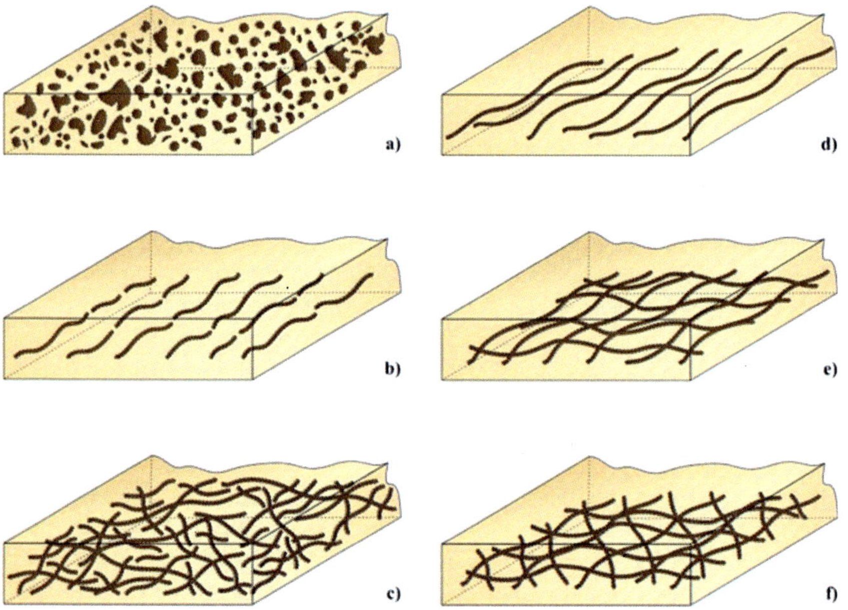

Fig. 1.1 Classification and types of composite materials. **a** Dispersed particle-reinforced, **b** Discontinuous fiber-reinforced (aligned), **c** Discontinuous fiber-reinforced (randomly oriented), **d** Continuous fiber-reinforced (aligned), **e** Continuous fiber-reinforced (aligned 0°–90° fiber orientation angle), and **f** Continuous fiber-reinforced (multidirectional fiber orientation angle)

the composite will receive during use. Mechanically speaking, matrix deformation must be compatible with the maximum deformation of the reinforcement [15–17].

Polymer composites can be defined as multi-phase materials formed by reinforcing fillers embedded in a polymer matrix.

Polymer matrices are classified as thermosets (epoxy, polyester, phenolic, silicone, polyimide, etc.) or thermoplastics (polyethylene, polystyrene, nylons, polycarbonate, polyether-ether ketone, polyphenylene sulfide, etc.). These matrices are most widely used due to their moderate cost, easy processing, good chemical resistance, and low density [1].

The use of conventional and engineering thermoplastics, as polymer matrices has been restricted to medium performance composites. The limited thermal stability of thermoplastics at high temperatures restricts their use for specific applications. However, the incorporation of reinforcements with specific characteristics in thermoplastic matrices has allowed applications of the polymer composite in temperatures up to 150 °C [15].

Epoxy resins, despite having excellent mechanical properties, are penalized by high costs and low resistance to weathering. Curing of these resins is much more complex than that of polyester resins. In the case of phenolic resins, their main disadvantage to the polyester and epoxy resins is that during their curing, water is formed as a by-product. Therefore, its application in composites is more complex, since the removal of produced moisture becomes an important factor during manufacturing [16].

Polyester resins are used in composites due to their low cost and satisfactory mechanical properties. After cured, these materials present good electrical properties, corrosion resistance and chemical attacks. Curing of the polyester resin is an exothermic process and requires an organic peroxide based curing system as the catalyst [16, 17]. Furthermore, polyester resins have ester groups as fundamental elements in their molecular chains, resulting from the condensation reaction of a diol with a diacid. Thus, depending on the type of acid used in the process, the polyester can be either saturated (thermoplastic) or unsaturated (thermoset) [17].

In general, fiber-reiforced polymer composite manufacturing techniques are classified as open mold and closed mold processes. The first allows for multi-directional orientation of the fibers relative to the mold or mandrel, except for pultrusion, where fibers are oriented in the direction parallel to the laminated surface. These processes include hand lay-up, spray up [18–22], filament winding [20, 21, 23], and pultrusion [20, 23, 24]. Closed mold process, in turn, produce fiber orientation in the direction parallel to the mold surface, includes compression molding and transfer molding [23, 25, 26].

Hand lay-up is the oldest and simplest manual production process; however, it also is the slowest. It can be adapted for the production of large structures, although it is more interesting for small productions [18].

Hand lay-up continues to be one of the most important manufacturing processes for small batch production of fiber reinforced composites, although increasing stringency of emission regulations has forced many manufacturers to explore the use of closed mold alternatives. Hand lay-up is carried out at room temperature using catalyzed liquid resins. The resin is poured in the mold containing the reinforcement. The matrix starts to cure in the mold, through an exothermic chemical reaction between the catalyst and the resin, solidifies and a fiber reinforced composite is obtained without the need of external heating. The method begins with the preparation of the mold where the reinforcing fiber layers will be stacked; usually a release agent is used on the mold, in order to facilitate demolding after curing. The impregnation process is performed manually with the help of rollers or brushes by pressing the resing and fiber layers to eliminate air bubbles and excess resin. Layers of fabric, mats, oriented or random fibers are stacked and impregnated with resin one by one in an open mold. After stacking all layers, an upper molding plate may be applied to allow for a better finish on both outer surfaces of the composite. Despite the steady progress to replace the manual layout with an automated process, it still persists as the method by which at least half of the entire advanced composite aerospace structures are made [19, 20, 27].

Hand lay-up continues to be used because it is extremely flexible to allow a wide variety of shapes of the desired composite [21]. Although it presents some disadvantages such as the quality of the final product depends on the ability of the worker, has low productivity and releases volatiles during manufacturing. Furthermore, as production volumes increase and economic factors become very important for decision making, manual placement gradually must be replaced or modified by the use of new automated technologies.

Nowadays, synthetic fiber-reinforced polymer composite materials are used in different sectors such as: aerospace, maritime, automotive, biomedical and transportation in specific applications as well as in high-tech sporting equipment [1–8, 12]. Already, vegetable fiber-reinforced polymer composites are used in the several construction sectors such as: automotive, building, transportation, consumer goods, sports equipment, design equipment, etc. [28–36].

Despite of the attractiveness, depending on the type of polymer composites, these materials are strongly affected when exposed to the adverse operating conditions such as electromagnetic and thermal radiation, galvanic corrosion (metals as reinforcement), oxygen at high temperatures (thermo-actived oxidative reactions), and water in the liquid and vapor (including moist air) phases. Therefore, innovative researches related to polymer composite degradation are strongly recommended, and with it, new challenges should arise.

References

1. Agarval, B., Broutman, L.J., Chandrashekha, K.: Analysis and Performance of Fiber Composites. John Wiley & Sons Inc., New Jersey (2006)
2. Brydson, J.A.: Plastics Materials, 7th edn. Butterworth-Heinemann, London (1999)
3. Crawford, R.J.: Plastics Engineering, 3rd edn. Butterworth-Heinemann, Oxford (1999)
4. Hull, D., Clyne, T.W.: An Introduction to Composite Materials, Cambridge Solid State Science Series, 2nd edn. Cambridge (1996)
5. Callister, W.D., Jr., Rethwisch, D.G.: Fundamental of Materials Science and Engineering: An Integrated Approach, 3rd edn. Jonh Wiley & Son Inc., USA (2008)
6. Peters, S.T.: Handbook of Composites. Chapman & Hall Cambridge University Press, England (1998)
7. Richardson, H.: Polymer Engineering Composites. Applied Science Publishers, London (1997)
8. Callister, W.D., Jr.: Materials Science and Engineering an Introduction. John Wiley & Sons Inc., New York (2007)
9. Shackelford, J.F.: Materials Science For Engineers, six Peason Prentice Hall, New Jersey (2005)
10. Flinn, R., Trojan, P.: Engineering Materials And Their Applications. Houghton Mifflin Company, Boston (1981)
11. Laranjeira, E.: Properties of Composites Polyester/Jute. Influence of the Addition of Nanoparticulate Mineral Charge and Antichama System. Thesis in Process Engineering. Federal University of Campina Grande (2004). Brazil, p 159 (In Portuguese)
12. Mitchell, B.S.: An Introduction to Materials Engineering and Science. John Wiley & Sons, Inc., New Jersey (2004)
13. Cavalcanti, W.S.: Composites Polyester/vegetable woven-glass: mechanical characterization and simulation of water sorption. Thesis in Process Engineering. Federal University of Campina Grande (2006). Brazil, p 122 (In Portuguese)
14. Kaw, A.K.: Mechanics Composite Materials, 2nd edn. Taylor & Francis Group Llc, New York (2006)
15. Hage Jr., E.: Polymeric Composites and Blends. Instituto Latino Americano e IBM, Campinas (1989) (In Portuguese)

16. Cheremisinoff, N.P.: Advanced Polymer Processing Operations. Noyes Publications, New Jersey (1998)
17. Goodman, S.H.: Handbook of Thermoset Plastics. Noyes Publications, New Jersey (1998)
18. Bunsell, A.R., Renard, J.: Fundamentals of Fibre Reinforced Composite Materials. Institute of Physics (IOP), UK (2005)
19. Astrom, B.T.: Manufacturing of Polymer Composites. CRC Press (1997)
20. Moura, M.F.S.F., Morais, A.B., Magalhães, A.G.: Composite Materials-Materials. Manufacturing and Mechanical Behavior. Publindústria, Porto (2010).(In Portuguese)
21. Gutowski, T.G.: Advanced Composites Manufacturing. Wiley, John & Sons, Cambrigde (1997)
22. Mazumdar, S.K.: Composites Manufacturing. Materials, Product and Process Engineering. Taylor & Francis, New York (2002)
23. Mallick, P.K.: Fiber-Reinforced Composites: Materials. Manufacturing and Design. CRC Press, New York (2007)
24. Huber, T., Graupner, N., Müssig, J.: Natural Fibre Composite Processing: A technical Overview. In: Mussig, J. (ed.) Industrial applications of natural fibres: structure, properties and technical applications, pp. 407–422. John Wiley & Sons (2010)
25. Ashby, M.F.: Materials and Design: Art and Science of Material Selection in Product Design. Elsevier, Rio de Janeiro (2011).(In Portuguese)
26. Ho, M.P., Wang, H., Lee, J.H., Ho, C.K., Lau, K.T., Leng, J., Hui, D.: Critical factors on manufacturing processes of natural fibre composites. Compos. B. Eng. **43**, 3549–3562 (2012)
27. Castro, B.F.M. : Study and Mechanical Characterization of Composites Reinforced with Natural Fibers, Master Dissertation in Mechanical Engineering, Porto (2013) (In Portuguese)
28. Dhakal, H.N., Zhang, Z.: The use of hemp as reinforcements in composites. In: Faruk, O., Sain, M. (Eds.) Biofiber Reinforcement in Composites Materials. Elsevier Ltda. , Amsterdam, Chapter 3, pp. 86–103 (2015)
29. Du, Y., Yan, N., Kortschot, M.T.: The Use Of Ramie Fibers as Reinforcements in Composites. In: Faruk, O., Sain, M. (Eds.) Biofiber Reinforcement in Composites Materials. Elsevier Ltda. , Amsterdam, Chapter 4, pp. 104–137 (2015)
30. Li, Y.; Shen, Y.O.: The Use of Sisal And Henequen Fibers as Reinforcements in Composites. In: Faruk, O., Sain, M. (Eds.) Biofiber Reinforcement in Composites Materials. Elsevier Ltda., Amsterdam, Chapter 6, pp. 165–210 (2015)
31. Leão, A.L., Cherian, B.M., Narine, S., Souza, S.F., Sain, M., Thomas, S.: The use of pineapple leaf fibers (palts) as reinforcements in composites. In: Faruk, O., Sain, M. (Eds.) Biofiber Reinforcement in Composites Materials. Elsevier Ltda. , Amsterdam, Chapter 7, pp. 211–235 (2015)
32. Mamun, A.A., Heim, H.P., Faruk, O., Bledzki, A.K.: The use of banana and abaca fibers as reinforcements in composites. In: Faruk, O., Sain, M. (Eds.) Biofiber Reinforcement in Composites Materials. Elsevier Ltda., Amsterdam, Chapter 8, pp. 236–272 (2015)
33. Kucak, D., Mistik, S.I.: The use of palm leaf fibers as reinforcements in composites. In: Faruk, O., Sain, M. (Eds.) Biofiber Reinforcement in Composites Materials. Elsevier Ltda., Amsterdam, Chapter 9, pp. 273–281 (2015)
34. Khalil, H.P.S.A., Alwani, M.S., Islam, M.N., Suhaily, S.S., Dungani, R., H'ng, Y.M.: The use of bamboo fibers as reinforcements in composites. In: Faruk, O., Sain, M. (Eds.) Biofiber Reinforcement in Composites Materials. Elsevier Ltda., Amsterdam, Chapter 16, pp. 488–524 (2015)
35. Shanks, R.A.: Isolation and application of cellulosic fibres in composites. In: Faruk, O., Sain, M. (Eds.) Biofiber Reinforcement In Composites Materials. Elsevier Ltda., Amsterdam, Chapter 18, pp.553–570 (2015)
36. Souza, S.F., Ferreira, M., Sain, M., Ferreira, M.Z., Pupo, H.F., Cherian, B.M., Leão, A.L.: The use of curaua fibers as reinforcements in composites. In: Faruk, O., Sain, M. (Eds.) Biofiber Reinforcement In Composites Materials. Elsevier Ltda., Amsterdam, Chapter 22, pp. 700–720 (2015)

Chapter 2
Fundamentals

This chapter focuses on the water diffusion process in composites materials. Special attention is given to vegetable fiber reinforced polymer composites. Herein, the main issues related to this important topic such as foundations, advantages and problems associated with the use of vegetable fibers as reinforcement in polymer composites are discussed, especially that caused by weak adhesion and exposure to the hot and humid environments, are presented and discussed.

2.1 Natural Fiber-Reinforced Polymer Composites

Currently, environmental issues are increasingly assuming a prominent role in our society, including conditioning technological development. In materials engineering, this role evidences the greater importance attributed to renewable and biodegradable materials whose manufacture involves low energy consumption. In this context, the evolution of different technologies has allowed the use of reinforced plastics with several productive, economic and ecological advantages.

More recently, environmental concerns as well as very stringent regulations and standards, have prompted the industries to research and use more eco-friendly and/or sustainable processing methods or products. In order to achieve this goal, available manufacturing processes are being modified or adapted to comply with these regulations, alternative technologies are developed and improved or else renewable biomaterials and biofibers are being used to generate eco-friendly composite materials. Broadly defined, biocomposites are composite materials made from natural/bio fibers; the polymer matrix may be petroleum based or naturally derived (biopolymer), and the former may be biodegradable or not. Biocomposites derived from plant fibers and biopolymers are likely to be more eco-friendly and such composites are termed green composites. While synthetic matrix/plant fiber composites are not as

J. M. P. Q. Delgado et al., *Moisture Transport in Polymer Composite Materials*, Advanced Structured Materials 160, https://doi.org/10.1007/978-3-030-77826-2_2

ecofriendly, they are still more environmental friendly than an all-synthetic material. Natural fibers, both of animal and plant (lignocellulosic) origin, meet these requirements.

Vegetable fibers are biodegradable, sustainable and renewable, carbon dioxide neutral and have a positive environmental impact, as they are nontoxic and nonabrasive to mixing and molding equipment, leading to significant production cost reductions. Their major constituents are cellulose, hemicellulose, lignin, pectin, waxes and water-soluble substances.

Plant fibers are a choice material for biocomposite production due to their worldwide availability, economical production, safer handling and working conditions compared to synthetic reinforcements, good set of mechanical properties, low cost per volume, low density and high relative strength and stiffness, which results in composites with specific strength and stiffness comparable to that of glass reinforced ones.

Strong arguments for encouraging the use of these materials include ecological, technological and economic advantages presented by vegetables fibers as well as social concerns since from an economical point of view, they generally originate from poor regions [1].

The main disadvantages presented by vegetable fibers for polymer composite applications are related to their high moisture sorption (hydrofilicity), relatively low processing temperatures ($\leq$200 °C (although it is possible to use higher temperatures for short periods), low resistance to microorganisms and poor fiber/polymer adhesion to most polymer matrices. All these factors ends up by compromising the composite´s immediate and long-term mechanical performance [2]. Therefore, despite their attractiveness, natural fiber-reinforced polymer matrix composites are very sensitive to influences from external environmental agents such as water (in the liquid or vapor phases), humid air, temperature, radiation (X-rays, β and γ rays, ultraviolet rays), chemical agents or any combination between them [3–5]. They are also very sensitive to internal factors such as fiber volume fraction and orientation, for example. The different factors cited above justify the significant amount of effort and research on understanding vegetable fiber-reinforced polymer composites degradation under adverse operating conditions. These actions also contemplate many others types of polymer composites, such as those reinforced by metallic fillers.

One of the most important factors in composite materials, manufactured by one or more continuous phases, is the region of interactions between the reinforcement and the matrix. It encompasses two sub-regions: the interface and the interphase. The interface is the boundary of contact (contact surface) between composite components (reinforcement and matrix) and has null thickness. In contrast, the interphase is a volumetric region where matrix and reirforcement are mixed and the chemical, physical and mechanical properties changes continuously from those of the matrix to reirforcement. Thus, its thickness is not null. The interphase is generated by the interdiffusion of atoms and molecules of the matrix across the interface, to inside the reinforcement. This phenomenon is more intense in situations involving polymer

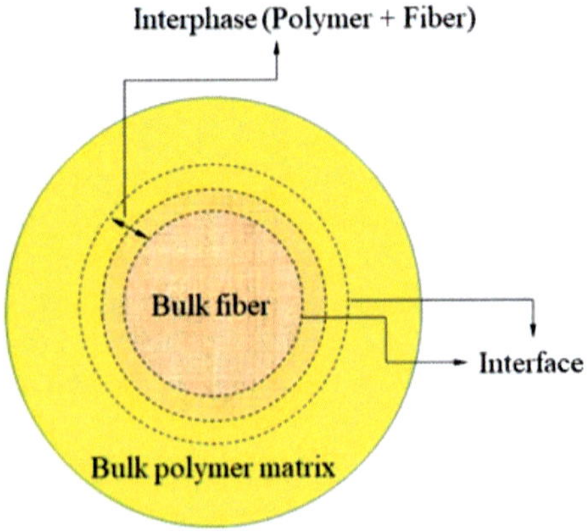

Fig. 2.1 A schematic illustration identifying the interface and interphase regions inside a polymer composite

matrices in the liquid state and vegetable fiber as reinforcement. Based on these definitions, two contact inferfaces constitute the interphase. Figure 2.1 illustrates the physical representations of the interface and interphase inside a fiber-reinforced polymer composite.

The final properties of the composites depend fundamentally on how the individual components interact with each other, that is, they depend on the interface and interphase regions between the discontinuous and continuous phases. These regions are mainly responsibles to transfer the mechanical load from the matrix to the reinforcement. Inadequate adhesion between the phases involved in the interface may lead to premature failures, compromising the performance of the composite. Therefore, in addition to the individual properties of each component of the composite, the interface should be as suitable as possible to optimize the combination of the properties involved [6–13]. The adhesion between reinforcement and matrix characterize the physical and mechanical behavior of a composite.

In polymer composites, the failure should occur in the matrix. In practice, the adhesion between the constituentes isn´t perfect and the process of rupture is generated at the interface. Therefore, in most cases, failure of reinforced polymers occurs by shearing in the interfacial region. The failure occurs due to the weakness of the atomic or intermolecular bonds between the surface of the matrix and the surface of the reinforcement in the composite. The following types of failures can be cited: matrix cracking, local deterioration, delamination, fiber debonding, and pull-out.

One of the factors that favor interfacial interaction is the contraction undergone by the polymer matrix during its curing (crosslinking and solidification). In order for a better stress distribution to occur on the reinforcement surface during matrix contraction, a perfect wetting of the reinforcement by the resin is necessary [9].

2.2 Water Absorption in Natural Fiber-Reinforced Polymer Composites

Despite their strongly hydrophobic characteristic, polymers can absorb water when immersed in aqueous media or exposed to moisture. The intensity of this phenomenon is dependent on factors such as: polymer polarity, hydrogen bonding ability, crystallinity (thermoplastic polymers), degree of crosslinking (thermoset polymers) and manufacturing process. Thus, moisture in any form is deleterious to polymer composites, especially to those reinforced by natural fibers, which are highly hydrophilic materials.

There are two modes of moisture absorption in vegetable fiber reinfroced polymer composites: (a) hydrogen bonds between the polymer and the hydrophilic groups of the filler and (b) through surface microcracks which are responsible for water transport and deposition inside the material [14].

The water absorbed by polymers consists of both free water and bound water [15]. The free water are water molecules with the ability to move independently through the void spaces, while bound water are water molecules that are attached to polar groups of the polymers [16].

Moisture diffusion in polymeric composites is governed by three different mechanisms. The first involves the diffusion of water molecules within micro gaps between polymer chains. The second involves moisture transfer through gaps and faults at the interface between fiber and matrix. This is a result of poor wetting and impregnation during the manufacturing process. The third involves the transport of water molecules through micro cracks in the matrix originated during manufacture [17–19].

When natural fiber-reinforced polymer composites are exposed to moisture, free water penetrates and binds with hydrophilic groups of the fiber, establishing intermolecular hydrogen bonds with the fiber and reducing interfacial adhesion between fiber and matrix. This leads to swelling of the cellulose fibers, which promotes an increase in the stress at the interfacial regions resulting in its embrittlement and thus, leading to formation of micro cracks inside the matrix around the swollen fibers [20–22]. This promotes capillarity and transport of moisture via micro cracks, causing deterioration of the fibers, which eventually lead to the definitive debonding between fibers and matrix. After long time periods, biological activities such as fungal growths eventually degrade the natural fibers [15, 16] compromising the mechanical performance of the composite. Thus, water absorption is one of the limiting factors that reduce the applicability, physical and mechanical properties of the composite [23]. Figure 2.2 illustrates the moisture types (Free water and bound water) and the effect of water migration on the fiber—matrix interface.

Considering that these materials may be exposed to moisture or even submerged in water during their use, water absorption studies are of great academic interest, especially when heated environments are involved [17, 24].

Vegetable fibers are hydrophilic and fiber moisture not only acts as plasticizer but makes polymer impregnation more difficult, causing weak adhesion on the polymer matrix-fiber interface, which leads to internal stresses, void formation and premature failure of the fiber-polymer system [25–27].

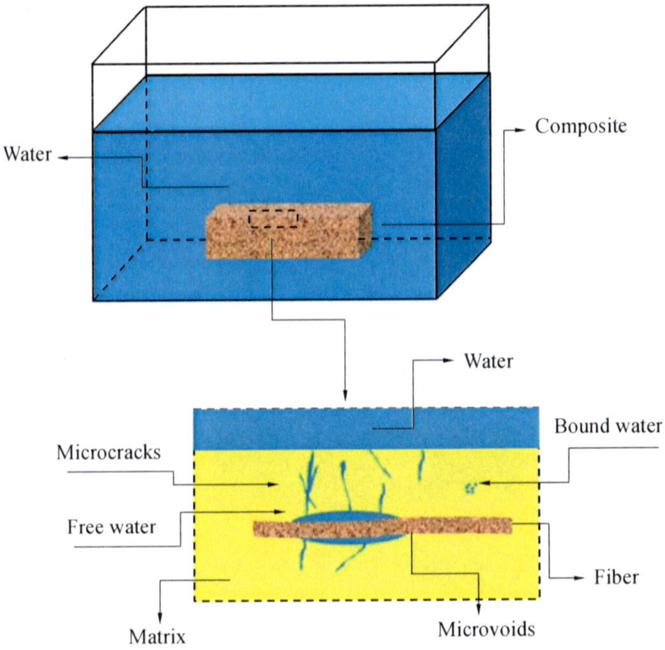

Fig. 2.2 Water migration mechanisms in vegetable fiber-reinforced polymer composite

This weak interfacial adhesion is associated with low polarity and low chemical affinity between the matrix and the fiber, which causes the formation of voids at the interface, initiation of faults and interfacial degradation that compromise the mechanical performance of the composites.

Effectively, vegetable fiber-reinforced polymer composites display lower mechanical properties than synthetic fiber-reinforced composites. However, as previously exposed, water sorption adversely affects the mechanical performance and physical integrity of natural fiber-reinforced polymer composite; so, the differences in mechanical properties between the natural and syntetic fiber-reinforced polymer composites can be even greater. Thus, knowing the effect of moisture on the composite properties is crucial for outdoor applications. The incompatibilities between natural fiber reinforced composites and water may be diminished by surface modification (chemical treatment) of the fiber or the matrix. In general it is the fiber, not the matrix, that is treated [28].

Several studies show that the mechanical properties of vegetable fiber reinforced composites significantly improve at high fiber content [29–31]. However, when fibrous polymer composites absorb moisture effects such as swelling, plasticizing, dissolving, leaching and/or hydrolyzing, result in discoloration, embrittlement, lower resistance to heat and weathering, and lower mechanical properties.

Due to their importance, several works on the water sorption kinetics of vegetable fiber reinforced polymer composites are reported in the literature [32–50]. In general, the amount of water absorbed by a sample varies as a function of its composition, dimensions, void fraction (available free volume), temperature, surface area, surface protection, and exposure time. In this context, the effects of moisture and temperature of composites on several performance parameters, such as tensile and shear strengths, elastic moduli, fatigue behavior, creep, rupture stress, response to dynamic impact, and electrical parameters (electrical resistance and dielectric constant, for example), have been regularly investigated [3].

References

1. Silva, R.V.: Composite polyurethane resin derived from castor oil and vegetable fibers. Thesis in Science and Materials Engineering. Federal University of São Paulo (2003). São Carlos, Brazil, 157f. (In Portuguese)
2. John, M.J., Thomas, S.: Biofibres and biocomposites. Carbohydr. Polym. **71**(3), 343–364 (2008)
3. Agarval, B., Broutman, L.J., Chandrashekha, K.: Analysis and Performance of Fiber Composites. John Wiley & Sons Inc., New Jersey (2006)
4. Callister, W.D., Jr.: Materials Science and Engineering an Introduction. John Wiley & Sons Inc., New York (2007)
5. Thwe, M.M., Liao, K.: Effects of environmental aging on the mechanical properties of bamboo-glass fiber reinforced polymer matrix hybrid composites. Compos. Part A Appl. Sci. Manuf. **33**, 43–52 (2002)
6. Callister Jr., W.D.: Materials Science and Engineering an Introduction, six.ed. John Wiley & Sons Inc., USA (2003)
7. Shackelford, J.F.: Materials Science For Engineers, six Peason Prentice Hall, New Jersey (2005)
8. Mitchell, B.S.: An Introduction to Materials Engineering and Science. John Wiley & Sons Inc., New Jersey (2004)
9. Cavalcanti, W.S.: Composites Polyester/vegetable woven-glass: mechanical characterization and simulation of water sorption. Thesis in Process Engineering. Federal University of Campina Grande (2006). Brazil, p 122 (In Portuguese)
10. Kaw, A.K.: Mechanics Composite Materials, 2nd edn. Taylor & Francis Group Llc, New York (2006)
11. Hull, D., Clyne, T.W.: An Introduction to Composite Materials, Cambridge Solid State Science Series, second ed. Cambridge (1996)
12. Hage Jr., E.: Compósitos e Blendas Poliméricas, Instituto Latino Americano e IBM, Campinas (1989)
13. Nóbrega, M.M.S.: Mechanical Properties of Polyester Composites Reinforced by Jute/Glass Hybrid Fabrics. Master Dissertation in Chemical Engineering, University Federal of Paraíba (2000). Campina Grande, Paraíba, Brazil (In Portuguese)
14. Sreekala, M.S., Kumaran, M.G., Thomas, S.: Water sorption in oil palm fiber reinforced phenol formaldeyde composites. Compos. Part A Appl. Sci. Manuf. **33**, 763–777 (2002)
15. Chen, H., Miao, M., Ding, X.: Influence of moisture absorption on the interfacial strength of bamboo/vinyl ester composites.Compos. Part A Appl Sci Manuf **40**(12), 2013–2019 (2009)
16. Azwa, Z.N., Yousif, B.F., Manalo, A.C., Karunasena, W.: A review on the degradability of polymeric composites based on natural fibres. Mater. Des. **47**, 424–442 (2013)
17. Espert, A., Vilaplana, F., Karlsson, S.: Comparison of water absorption in natural cellulosic fiber from wood and one-year crops in polypropylene composites and its influence on their mechanical properties. Compos. A Appl. Sci. Manuf. **35**(11), 1267–1276 (2004)
18. Lin, Q., Zhou, X., Dai, G.: Effect of hydrothermal environment on moisture absorption and mechanical properties of wood flour-filled polypropylene composites. J. Appl. Polym. Sci. **85**(14), 2824–2832 (2002)

19. Comyn, J.: Polymer Permeability. Elsevier, London, England, 383 (1985)
20. Li, Y., Mai, Y.W., Ye, L.: Sisal fibre and its composites: a review of recent developments. Compos. Sci. Technol. **60**(11), 2037–2055 (2000)
21. Wambua, P., Ivens, J., Verpoest, I.: Natural fibres: can they replace glass in fibre reinforced plastics? Compos. Sci. Technol. **63**(9), 1259–1264 (2003)
22. Aziz, S.H., Ansell, M.P., Clarke, S.J., Panteny, S.R.: Modified polyester resins for natural fibre composites. Compos. Sci. Technol. **65**(3–4), 525–535 (2005)
23. Hodzic, A., Shanks, R.: Natural fibre composites-materials, processes and properties. Ed. Woodhead Publishing Limited, Philadelphia, USA (2014)
24. Pothan, L., Thomas, S.: Effect of hybridization and chemical modification on the water-absorption behaviour of banana fiber-reinforced polyester composites. J. Appl. Polym. Sci. **91**(6), 3856–3865 (2004)
25. Carvalho, L.H., Canedo, E.L., Neto, S.F., Lima, A.G.B.: Moisture transport process in vegetable fiber composites: theory and analysis for technological applications. In: Delgado, J.M.P.Q. (ed.) Industrial and Technological Applications of Transport in Porous Materials, pp. 37–62. Berlin Heidelberg, Springer (2013)
26. Kim, J.-K., Mai, Y.-W.: Engineered Interfaces in Fiber Reinforced Composites. Elsevier, Amsterdam (1998)
27. Bunsell, A.R., Renard, J.: Fundamentals of Fibre Reinforced Composite Materials. IOP Publishing Ltda, Bristol (2005)
28. Li, X., Tabil, L.G., Panigrahi, S.: Chemical treatments of natural fiber for use in natural fiber-reinforced composites: a review. J. Polym. Environ. **15**, 25–33 (2007)
29. Joseph, P.V., Rabello, M.S., Mattoso, L.H.C., Joseph, K., Thomas, S.: Environmental effects on the degradation behaviour of sisal fibre reinforced polypropylene composites Compos. Sci. Technol. **62**, 1357–1372 (2002)
30. Wessler, K., Fogagnolo, C., Everling, M., Bernardo, H.P., Sobral, J.C., Balzer, P. S., Araujo, M.I.S.: Obtenção e caracterização de compósitos de resina poliéster e fibras de bananeira. In: Congresso de Ciência dos Materiais do Mercosul SULMAT 2004, Joinvile, Brazil, 1–6 (2004)
31. Carvalho, L.H., Ladchumananandasivam, R., Alexandre, M.E.O., Cavalcanti, W.S.: Propriedades de compósitos poliéster/fibra da folha do abacaxi. In: Congresso em Ciência dos Materiais do Mercosul. SULMAT 2004, Joinvile, Brazil, 1–6 (2004)
32. Kucak, D., Mistik, S.I.: The use of palm leaf fibers as reinforcements in composites. In: Faruk, O., Sain, M. (Eds.) Biofiber reinforcement in composites materials. Elsevier Ltda. , Amsterdam, Chapter 9, pp. 273–281
33. Rouison, D., Couturier, M., Sain, M., Macmillan, B., Balcom, B.J.: Water absorption of hemp fiber/unsaturated polyester composites. Polym. Compos. **26**(4), 509–524 (2005)
34. Dhakal, H.N., Zhang, Z.: The use of hemp as reinforcements in composites. In: Faruk, O., Sain, M. (Eds.) Biofiber Reinforcement In Composites Materials. Elsevier Ltda., Amsterdam, Chapter 3, pp. 86–103 (2007)
35. Dhakal, H.N., Zhang, Z.Y., Richardson M.O.W.: Effect of water absorption on the mechanical properties of hemp fibre reinforced unsaturated polyester composites. Compos. Sci. Technol. **67**(7–8), 1674–1683 (2007)
36. Bismarck, A., Askargorta, I.A., Springer, J., Lampke, T., Wielage, B., Stamboulis, A.: Surface characterization of flax, hemp and cellulose fibres; surface properties and the water uptake behaviour. Polym. Compos. **23**(5), 872–894 (2002)
37. Paiva, J.M.F., Frollini, E.: Phenolic thermofix matrix in reinforced composites with sugarcane bagasse fibers. Polímeros. **2**, 78–87 (1999)
38. Sanchez, E.M., Cavani, C.S., Leal, C.V., Sanchez, C.G.: Composites of unsaturated polyester resin with sugarcane bagasse: influence of fiber treatment on properties. Polymer **20**(3), 194–200 (2010)
39. Du, Y., Yan, N., Kortschot, M.T.: The use of ramie fibers as reinforcements in composites. In: Faruk, O., Sain, M. (Eds.) Biofiber Reinforcement in Composites Materials. Elsevier Ltda., Amsterdam, Chapter 4, pp. 104–137 (2015)

40. Leão, A.L., Cherian, B.M., Narine, S., Souza, S.F., Sain, M., Thomas, S.: The use of pineapple leaf fibers (palts) as reinforcements in composites. In: Faruk, O., Sain, M. (Eds.) Biofiber Reinforcement in Composites Materials. Elsevier Ltda., Amsterdam, Chapter 7, pp. 211–235 (2015)
41. Jayamol, G., Bhagawan, S.S., Thomas, S.: Effects of environment on the properties of low-density polyethylene composites reinforced with pineapple-leaf fibre. Compos. Sci. Technol. **58**(9), 1471–1485 (1998)
42. Khalil, H.P.S.A., Alwani, M.S., Islam, M.N., Suhaily, S.S., Dungani, R., H'ng, Y.M.: The use of bamboo fibers as reinforcements in composites. In: Faruk, O., Sain, M. (Eds.) Biofiber Reinforcement in Composites Materials. Elsevier Ltda., Amsterdam, Chapter 16, pp. 488–524 (2015)
43. Liu, W., Qiu, R., Li, K.: Effects of fiber extraction, morphology, and surface modification on the mechanical properties and water absorption of bamboo fibers-unsaturated polyester composites. Polym. Compos. **37**(5), 1612–1619 (2016)
44. Souza, S.F., Ferreira, M., Sain, M., Ferreira, M.Z., Pupo, H.F., Cherian, B.M., Leão, A.L.: The use of curaua fibers as reinforcements in composites. In: Faruk, O., Sain, M. (Eds.) Biofiber Reinforcement In Composites Materials. Elsevier Ltda., Amsterdam, Chapter 22, pp. 700–720 (2015)
45. Mamun, A.A., Heim, H.P., Faruk, O., Bledzki, A.K.: The use of banana and abaca fibers as reinforcements in composites. In: Faruk, O., Sain, M. (Eds.) Biofiber Reinforcement In Composites Materials. Elsevier Ltda., Amsterdam, Chapter 8, pp. 236–272 (2015)
46. Haneefa, A., Bindu, P., Arvind, I., Thomas, S.: Studies on tensile and flexural properties of short banana/glass hybrid fiber reinforced polystyrene composites. J. Compos. Mater. **42**(15), 1471–1489 (2008)
47. Kiran, C.U., Reddy, G.R., Dabade, B.M., Rajesham, S.: Tensile properties of sun hemp, banana and sisal fiber reinforced with polyester composites. J. Reinf. Plast. Compos. **26**(10), 1043–1050 (2007)
48. Badia, J.D., Kittikorn, T., Strömberg, E., Santonja-Blasco, L., Martínez-Felipe, A., Ribes-Greus, A., Ek, M., Karlsson, S.: Water absorption and hydrothermal performance of PHBV/sisal biocomposites. Polym. Degrad. Stab **108**, 166–174 (2014)
49. Chow, C.P.L., Xing, X.S., Li, R.K.Y.: Moisture absorption studies of sisal fibre reinforced polypropylene composites Compos. Sci. Technol. **67**, 306–313 (2007)
50. Li, Y., Shen, Y.O.: The use of sisal and henequen fibers as reinforcements in composites. In: Faruk, O., Sain, M. (Eds.) Biofiber Reinforcement in Composites Materials. Elsevier Ltda., Amsterdam, Chapter 6, pp. 165–210 (2015)

Chapter 3
Experimental Analysis

This chapter provides experimental information on water absorption in unsaturated porous media, with particular reference to vegetable fiber-reinforced polymer composites. Herein, important information related to fiber morphology, manufacturing, moisture absorption, and mechanical caracterization tests in polymer composites reinforced by caroá, macambira and sisal fibers are presented in detail. Results of the fiber morphology, moisture gain kinetics, for different sample thickness and process temperature, and mechanical properties (tensile strength, elastic modulus, impact resistance and elongation) are shown and analyzed.

3.1 Background

In physical problems involving moisture absorption or desorption in composites, it is very important to determine the moisture content at any instant of the process. In general, mass transfer rate from or to the body depends on different factors such as temperature, filler content, initial moisture content, and nature and orientation of the reinforcement. Further, hydric, thermal and mechanical events (in a micro scale analysis) occurring in short fiber reinforced composites are very different from those verified for long fiber reinforced composites [1].

All of the factors mentioned above affect mechanical performance and properties and limit the field of application of vegetable reinforced polymer composites [2–9] and new researches in this area are required. Several of the papers reported in the literature are directed to the problem of moisture absorption by composites reinforced by plant fibers and their influence on the properties of the material such as those by Marcovich et al. [10], Thwe and Liao [11], Mulinari [12], Sreekala et al. [13], Espert et al. [14], Angrizani et al. [15], Fernandes [16], Vieira [17], Jayamol et al. [18], Cavalcanti et al. [19], Santos [20], Sanchez et al. [21], Tita et al. [22], Rao et al. [23],

J. M. P. Q. Delgado et al., *Moisture Transport in Polymer Composite Materials*,
Advanced Structured Materials 160, https://doi.org/10.1007/978-3-030-77826-2_3

Sensarzadeh and Amiri [24], Idriss et al. [25], Soni and Soni [26], Srihari et al. [27], Pavlidou and Papaspyrides [28], Pegoretti and Penati [29], Kumosa et el. [30], and many others.

As a complement to the theme discussed in this book, and due to the importance of accurately describing the water sorption phenomena in vegetable fiber-reinforced polymer composites, experimental studies about water sorption by unsaturated polyester composites reinforced with short fibers of macambira (*Bromelia Laciniosa*), caroá (*Neoglaziovia Variegata*), and sisal (*Agave sisalana*) fibers are presented here.

3.2 Experimental Setup

Nóbrega [8] and Nóbrega et al. [31] conducted several experiments on the water absorption of caroá fiber-reinforced unsaturated polyester composites. Similarly Cruz et al. [32] conducted experiments with macambira fiber-reinforced unsaturated polyester composites, while Santos [20] conducted sorption experiments with sisal fiber-reinforced unsaturated polyester composites.

Macambira fibers (Fig. 3.1) are extracted from the macambira plant, which belongs to the family Bromeliaceae. The plant used in the study, was obtained from the Cariri region of the State of Paraíba State in Brazil and its chemical composition was determined to be: α-cellulose (58.72%), hemicellulose (19.37%), lignin (12.62%), moisture (8.94%), and others [33].

Caroá fibers (Fig. 3.2) are extracted from the caroá plant, which also belongs to the family Bromeliaceae. The plant employed in the study reported here was obtained at a farm in the town of Pocinhos in the State of Paraíba State in Brazil. Caroá leaves provide long fibers, of great resistance and durability. The chemical composition of the caroá fiber is as follows: cellulose (35.5%), hemicellulose (17.9%) and lignin (30.1%) and others [8, 34].

(a) (b)

Fig. 3.1 Macambira (Bromelia Laciniosa). **a** Plant and **b** fibers

(a) (b)

Fig. 3.2 Caroá (Neoglaziovia Variegata). **a** Plant and **b** fibers

Sisal fibers (Fig. 3.3) are extracted from the Sisal plant which belongs to the family Agavaceae. The plant obtained in the study reported here came from a farm in the town of Pocinhos in the State of Paraíba in Brazil. Sisal leaves provide long fibers, of great resistance and durability. The chemical composition of the sisal fiber is as follows: cellulose (65.8%), hemicellulose (12.0%) and lignin (9.9%) and others [6, 20, 34].

The fibers of caroá and macambira were washed with running water, allowed to air dry at room temperature for at least 72 h before being combed and cut up to 5 cm.

(b)

(a)

Fig. 3.3 Sisal (Agave sisalana). **a** Plant and **b** fibers

The sisal fibers were dried in an air circulation oven operating at 60 °C until constant weight and subsequently heated at 105 °C for 24 h, in order to obtain the dry fiber. The dried sisal fibers were cut in 35 mm lengths prior to use. Unsaturated polyester cured with 1% MEK (methyl ethyl ketone) was used as the matrix.

Composites with varying fiber contents were compression molded. Since the thermal degradation of natural fibers and thermoset or thermoplastic polymers occurs at upper temperature limits, composite samples were made at low temperature. A fiber mat was produced by randomly distributing a pre-determined amount of fibers in steel molds (220 × 180 × 3 mm and 220 × 180 × 6 mm) and compressed with 2 ton for 5 min at room temperature. The mats thus produced were removed from the mold for further use as reinforcement. An appropriate quantity of polyester resin was mixed with the catalyst and a small amount was poured onto the mold. The fiber mats were placed in the steel mold, impregnated with more resin and the mold was closed (8 ton).

The system was allowed to cure under pressure for 4 h at room temperature before the composite plate was removed from the mold. The caroá and macabira reinforced composite plates obtained were post-cured in an air circulating oven operating at 50 °C for 48 h (Fig. 3.4), while sisal composite plates were post-cured in an air circulating oven operating at 60 °C for 48 h.

(a)

(b)

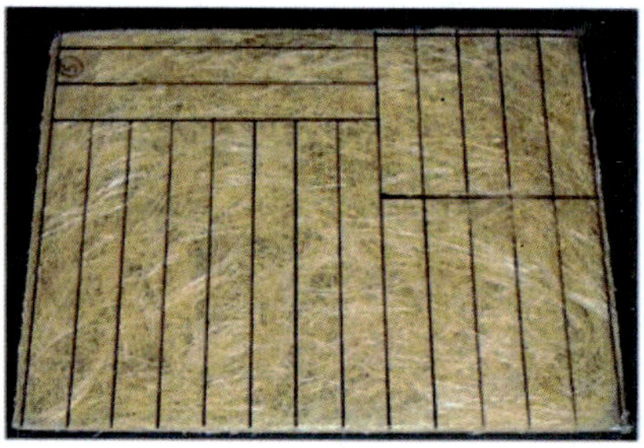

(c)

Fig. 3.4 Manufacturing of caroá fiber-reinforced polymer composite by hand lay up technique. **a** Fiber mat in the mold, **b** Demolding, and **c** cut samples

Tests for mechanical properties were conducted according to ASTM standards D-3039 and D-256 for tensile and impact tests (Fig. 3.5a–b), respectively. Composites samples of 20 × 20 mm were cut-off from the plates, and their edges were sealed with resin prior to the water absorption tests (to avoid water transport by capillarity) and dried in an air-circulating oven at 105 °C to constant weight or dry mass (Fig. 3.5c).

The water absorption experiments were carried out according to the following procedure. Firstly, the pre-dried composites samples were fully immersed in water baths (Fig. 3.6) kept at 25, 50 and 70 °C. At regular time intervals, the samples were removed from the water bath; pat-dried with a paper tissue to remove surface water

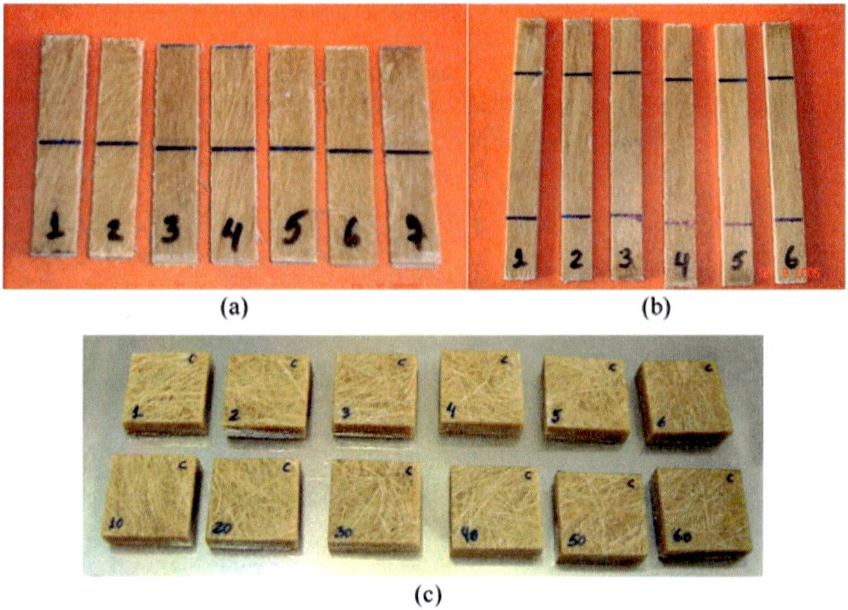

Fig. 3.5 Caroá fiber-reinforced polymer composite samples used at different test. **a** Impact, **b** Tensile and **c** Water sorption

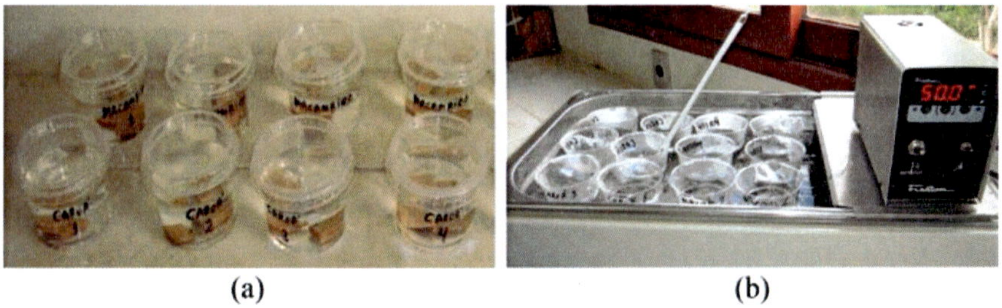

Fig. 3.6 **a** Caroá and macambira composite specimens and **b** Macambira and caroá composites samples in water bath

and immediately weighted in a digital scale (uncertainty ±0.001 g). The samples were re-immersed in the water bath and the procedure repeated so that the water sorption process continued until equilibrium was reached. Each measurement took less than 1 min, so water evaporation at the surface was deemed insignificant.

The results of absorbed moisture were reported as mass of absorbed water per unit of dry composites mass. Thus, moisture content was computed as follows:

$$\bar{M}(t) = \frac{m_t - m_0}{m_0} \times 100\% \tag{3.1}$$

where m_0 and m_t represent the dry mass of the composites samples ($t = 0$) and the wet mass at any specific time t, respectively.

Saturation (equilibrium) condition was assumed when the daily weight gain of the composite samples was less than 0.1%. We notice that complete immersion of composite samples in water bath constitutes the most severe physical situation; exposure to humid air results in lower equilibrium moisture content.

3.3 Experimental Results Analysis

3.3.1 Morphology Characterization of the Fibers

The microstructure of vegetable fibers is complex, and dependent on each kind of the fiber. In this research some single caroá, macambira and sisal fiber samples (untreated) were analyzed by scanning electron microscopy (SEM), in order to characterize their morphology. Figures 3.7, 3.8 and 3.9 present micrographs (transversal section) of macambira, caroá and sisal fibers in the natural state, respectively.

Analysis of these figures shows that the arrangement is similar to other natural fibers, with a spongy aspect and thin, compacted cellular arrangement (regular fiblilar arrangement). In some places is verified the presence of small amout of organic material at the fiber surface, probably a residue left when the fibers were extracted from the leaves of the each plant.

Images obtained from SEM also showed rougher surfaces and many void spaces, which implies that macambira, caroá ans sisal fibers could be adequate to be used as reinforcement in composite materials; this surface condition allows good adhesion between fibers and polymer matrix. However, it is important to state that, after treatment or mechanical characterization, the fibers achieve some degradation, and impurities existing at the surface are minimized. Furthermore, since cellulose is the major constituent with a crystalline structure, it is the main contributor towards the mechanical properties.

From the viewpoint of water absorption, cellulose fibers are difficult to dissolve because of their high crystallinity, but, they tend to retain liquids in the interfibrilar space [35].

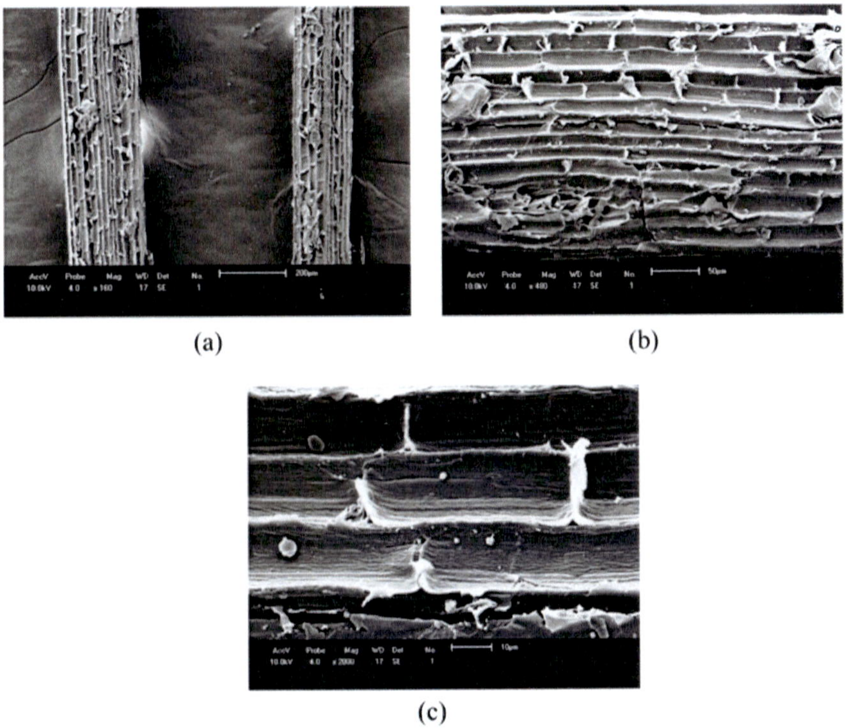

Fig. 3.7 SEM micrographs for untreated macambira fiber: **a** 160×, **b** 480× and **b** 2000×

3.3.2 Water Absorption Kinetics

Water sorption curves of unsaturated polyester/macambira composites, unsaturated polyester/caroá composites and unsaturated polyester/sisal composites are illustrated in Figs. 3.10 and 3.11. Results are reported for different dimensions of the samples (3 mm and 6 mm thickness) and water bath temperatures (25, 50 and 70 °C) as function of exposure (immersion) time. Upon analyzing of these figures it can be seen that the general shape of the curves is similar to those of others natural fiber-reinforced polymer matrix composites.

For all investigated composites the moisture content increases monotonically with water immersion time until it reaches a maximum value, the so-called equilibrium moisture content (saturation or hygroscopic condition). This behavior strongly indicates that the vegetable fibers were uniformly distributed in the matrix, as reported in the literature [2]. Figures 3.10, 3.11 and 3.12 also show that water sorption for all composites increase with the area/volume ratio and temperature (Table 3.1). Sample size effect was more pronounced at the lower temperature.

Table 3.1 summarizes the saturation data for all tested composites and for neat polyester. The data indicates that water absorption for the composites is higher than

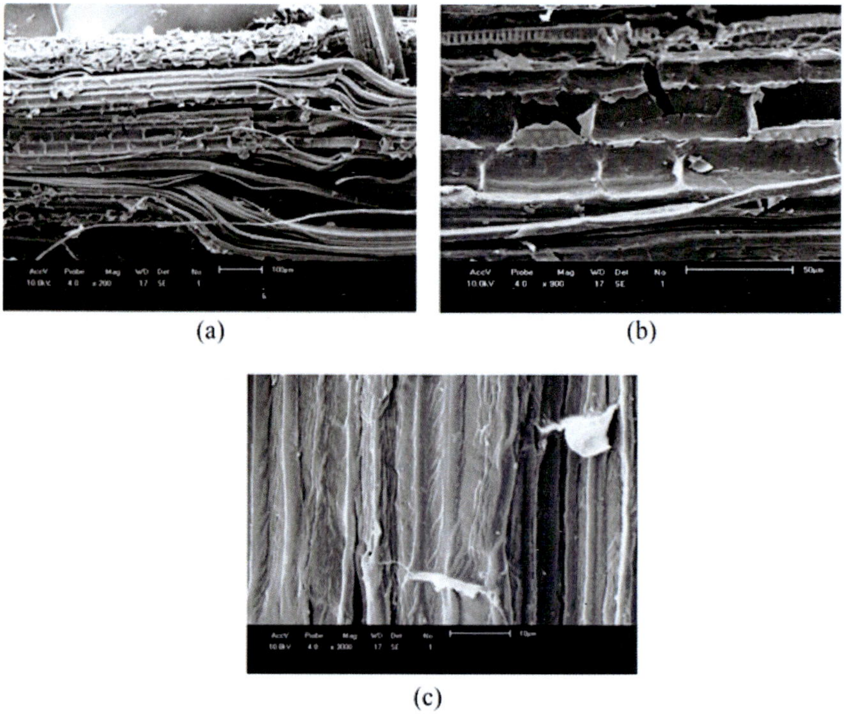

Fig. 3.8 SEM micrographs for untreated caroá fiber: **a** 200×, **b** 900× and **b** 3000×

for the matrix. Neat polyester shows a very low percentage of absorbed water. While the water sorption at the equilibrium condition for the unsaturated polyester is about 1%, for the composites it ranged from 14.04 to 18.09%. It was verified that the weight gain of pure polyester was almost insignificant, indicating minimal degradation. These results are consistent with studies reported in the literature for similar systems [2, 8, 10, 13, 36]. Some authors attribute the increase in the water sorption by polymer composites reinforced with vegetable fiber not only to the hydrophilic nature and the permeability of this type of reinforcement but also to the sample surface area exposed to water. Capillarity effects and the interfacial area between fiber and matrix may also contribute to water pick-up by these composites.

Results indicate that the water uptake of the composite sample immersed in a water bath at 70 °C was faster than under the other (25 and 50 °C) experimental conditions. This behavior is attributed to the increased water mobility within the solid at higher temperature. It is believed that higher water temperatures lead to thermal dilation of the composites and to increased composite porosity, which would, in turn, cause a faster moisture migration. Temperature activates the water diffusion process inside the sample, and sorption rate increases with the increases in temperature (thermo-activation).

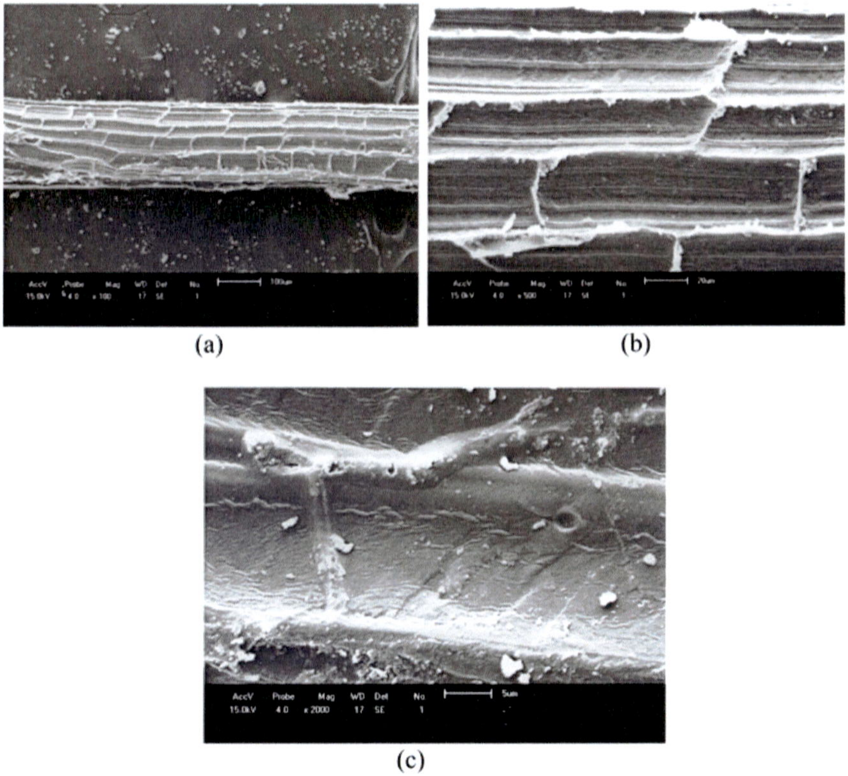

Fig. 3.9 SEM micrographs for untreated sisal fiber: **a** 100×, **b** 500× and **b** 2000×

These results are consistent with those reported in similar systems for unsaturated polyester/caroá composites [8, 31], unsaturated polyester/jute composites [36] and unsaturated polypropylene/sisal composites [35]. Other aspects were not analyzed, such as fiber size and orientation (events that occur in long fiber reinforced composites are different those observed in short fiber reinforced composites) [1], and water migration by capillarity into micro cracks inside the solid, mainly in the fiber-matrix interface where adhesion plays an important role. Reading reported works on these topics is strongly recommended.

3.3.3 Mechanical Properties

As already mentioned in the previous sections, the mechanical performance of the composites depends on the properties of both reinforcement and matrix, and the adhesion between them after manufacturing. Figure 3.13 show the results of tensile strength of the caroá fiber-reinforced polymer composites (in dry condition) as a

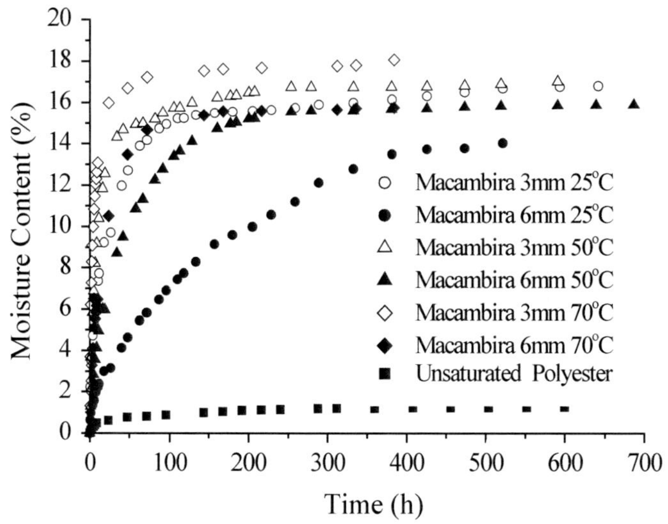

Fig. 3.10 Water sorption kinetics of macambira fiber-reinforced unsaturated polyester composites (30% w/w fiber content)

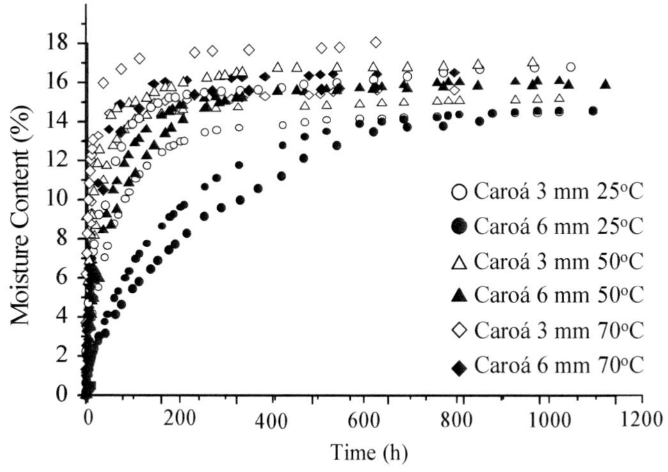

Fig. 3.11 Water sorption kinetics of caroá fiber-reinforced unsaturated polyester composites (30% w/w fiber content)

function of fiber content. Analysis of this figure, shows that the tensile strength of the composite is lower than that of the matrix for loadings up to 25% by weight for caroá fibers. This behavior can be attributed to inefficient loading, so the fibers act as defects (stress concentrators) and effectively weaken the matrix, resulting in a composite with lower mechanical strength. At low fiber content, the matrix is not sufficiently fixed

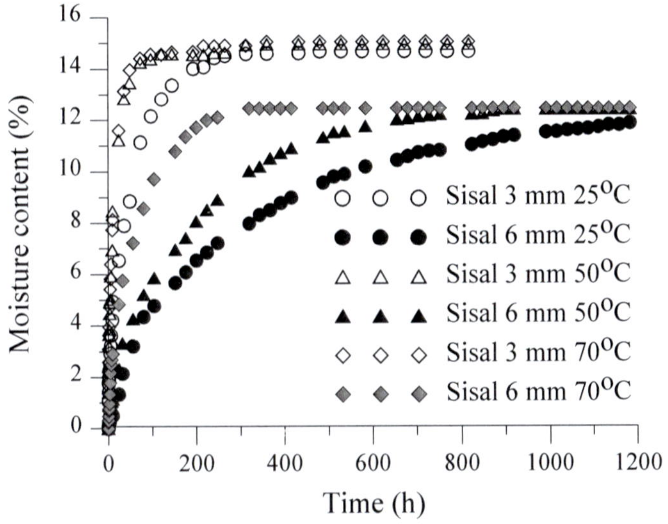

Fig. 3.12 Water sorption kinetics of sisal fiber-reinforced unsaturated polyester composites (46.6% w/w fiber content)

(poor adhesion) and high deformations imposed on it lead to a break of the matrix-fiber bond. As the loading increases, stresses are more uniformly distributed and the composite's strength increases. Obviously this behavior is dependent on the critical volume of the fiber into the matrix.

At loadings above 40% by weight, the tensile strength of the composites decreases with a further increase in fiber content. This behavior associated with fiber agglomeration and poor impregnation, which leads to an increase in their effective diameter, a decrease in their aspect ratio and to void formation.

Figure 3.14 illustrates the effect of the fiber content on the elastic modulus of the caroá fiber-reinforced polymer composite composites (in dry condition). The results show that fiber incorporation increases the stiffness of the composite. This behavior was expected, since the strength and Young´s modulus of the fibers used are higher than the matrix. In addition, fiber addition minimizes the movement of polymer chains resulting in an increase in the elasticity modulus of the composite with fiber content.

Figure 3.15 displays the impact resistance of the caroá fiber-reinforced polymer composite as a function of fiber content (in dry condition). Upon analyzing this figure, it can verified that impact strength increases with fiber content at all loading levels for the type of fiber tested. Based on the results, we state that the caroá fiber is able to diverge crack propagation and delay breakage, thus increasing impact strength. It is worth noticing that impact properties are not as dependent on fiber/matrix adhesion as the tensile strength and, in fact, this mechanical property often is improved for physical situations with looser interfaces.

Table 3.1 Moisture content and geometric data for the water sorption in vegetable fiber reinforced composites (30% w/w fiber contents)

Sample	Thickness (mm)	Tw (°C)	t (h)	M_t (%) (d. b.) ($t \to \infty$)	Area/Volume (mm^2/mm^3) (t = 0)
Polyester matrix	3.00	25	600	1.26	866.67
Macambira composite	3.00	25	642	16.81	866.67
	6.00	25	594	14.04	533.33
	3.00	50	687	17.02	866.67
	6.00	50	687	15.89	533.33
	3.00	70	384	18.09	866.67
	6.00	70	648	16.18	533.33
Caroá composite	3.00	25	1100	14.49	866.67
	6.00	25	1623	14.81	533.33
	3.00	50	1022	15.16	866.67
	6.00	50	1022	16.08	533.33
	3.00	70	796	15.61	866.67
	6.00	70	796	16.52	533.33
Sisal composite	3.00	25	817	14.68	533.33
	6.00	25	1328	12.27	866.67
	3.00	50	817	14.96	533.33
	6.00	50	848	12.42	866.67
	3.00	70	817	15.05	533.33
	6.00	70	848	12.46	866.67

Figure 3.16 shows the elongation at break of the caroá fiber-reinforced polyester composite as function of fiber content (in dry condition). It can be observed that elongation at break increases with fiber incorporation the incorporation of fiber increases the properties of elongation at fiber contents above 23%. This behavior is attributed to the greater stiffness of the fiber, thus, increasing the stiffness of the composite.

Since moisture absorption affects the mechanical performance of fiber-reinforced polymer composites, results of the mechanical tests of sisal fiber-reinforced polymer composites are shown in Figs. 3.17, 3.18, 3.19 and 3.20, as reported by Santos [20] and Santos et al. [37]. These results were obtained with 3 and 6 mm thick samples. Saturation conditions at different water bath temperatures are cited in Table 3.1.

Analysis of these figures indicate that tensile strength, impact strength, Young's modulus and elongation at break of the composites decrease with the increasing water bath temperature. This behavior can probably be associated with the fact that an increase in water bath temperature provokes deterioration and dilation of the samples and, thus, increasing moisture rate and facilitating water migration to inside the composite and, thus leading to a decrease in mechanical properties. The effect of sample thickness also is evident.

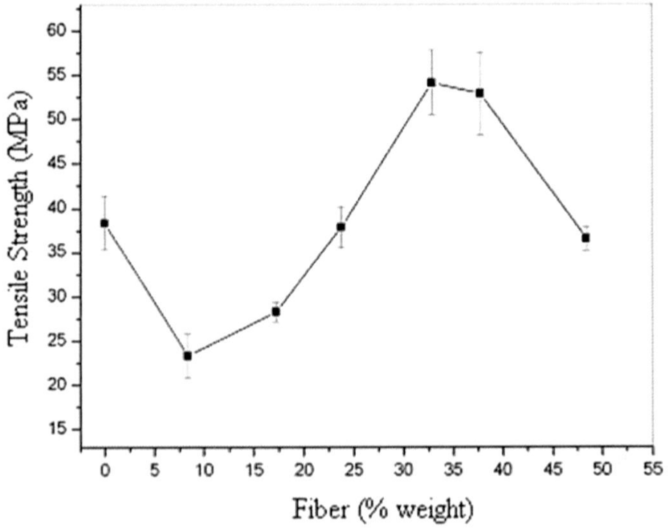

Fig. 3.13 Tensile strength of the caroá fiber-reinforced polymer composite as a function of the fiber content

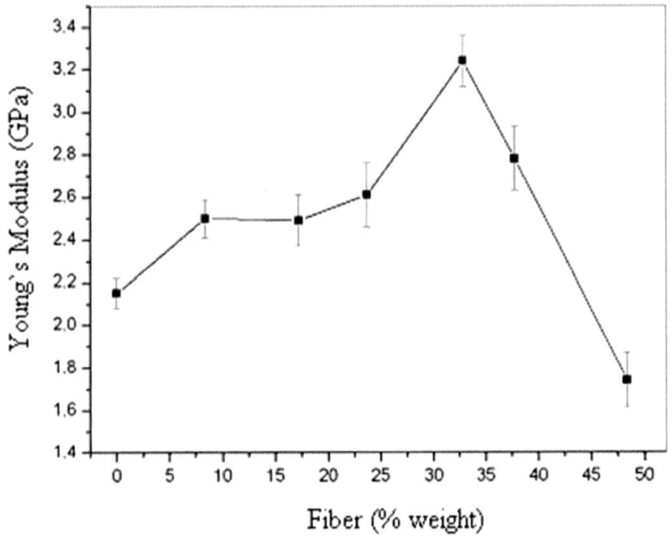

Fig. 3.14 Young's modulus of caroá fiber-reinforced polymer composite as a function of the fiber content

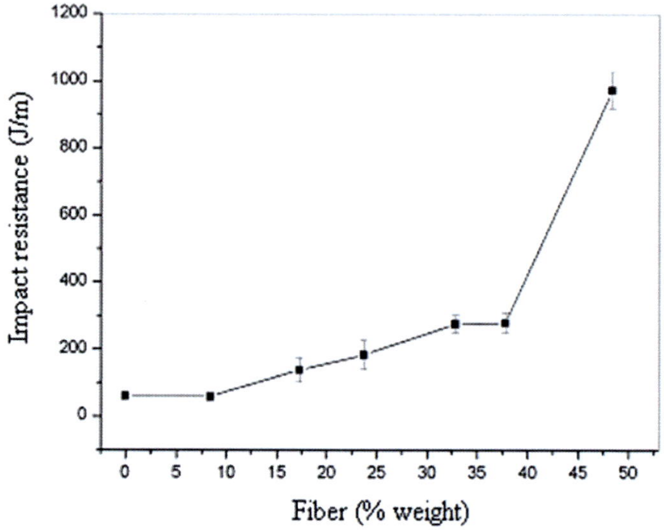

Fig. 3.15 Impact resistance of caroá fiber-reinforced polymer composite as a function of the fiber content

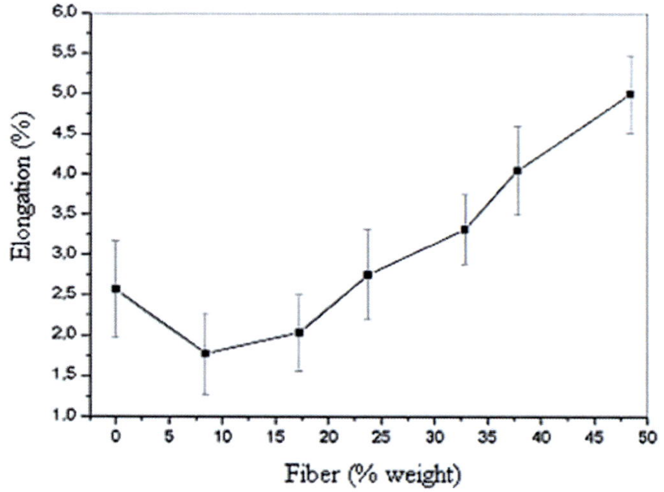

Fig. 3.16 Elongation at break of caroá fiber-reinforced polymer composite as a function of fiber content

Furthermore, it is important to notice that the decrease in mechanical property values of the composite are associated with the poor fiber-matrix interfacial adhesion caused by water absorption as previously discussed [14].

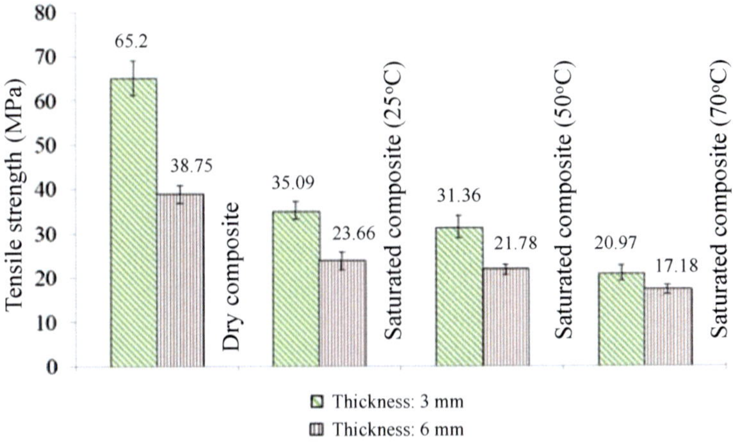

Fig. 3.17 Tensile strength of polyester composite (44.6% sisal fiber)

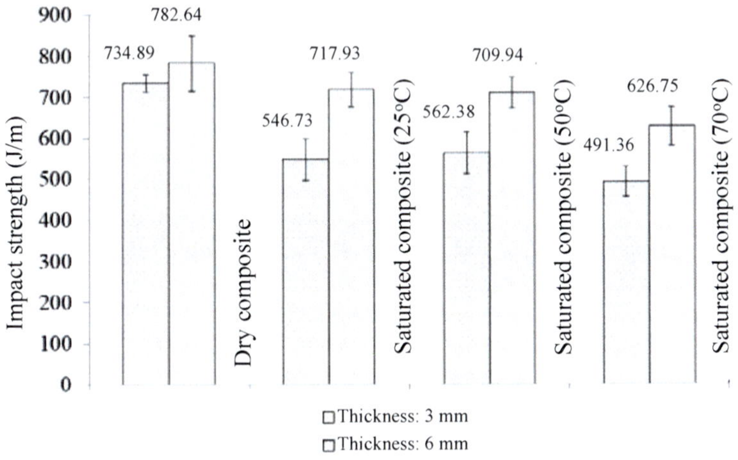

Fig. 3.18 Impact resistance of polyester composite (44.6% sisal fiber)

Different methods to improve fiber/matrix interfacial adhesion have been used in order to minimize this problem. In general, these methods are based on chemical treatment or by addition of coupling agent, as reported by Du et al. [38]. However, drying of vegetable fibers as a pre-treatment has been used as well [39]. Carvalho et al. [39] reported an experimental study about tensile properties of hybrid jute/cotton and sisal/cotton fabrics–polyester matrix composites. According to the authors, the amount of absorbed water on the composites was effectively reduced when the fabrics were pre-dried before their incorporation into the composites. Furthermore, it was verified that composites manufactured with pre-dried fabrics showed smaller strain at break, indicating a better fabric-matrix interaction.

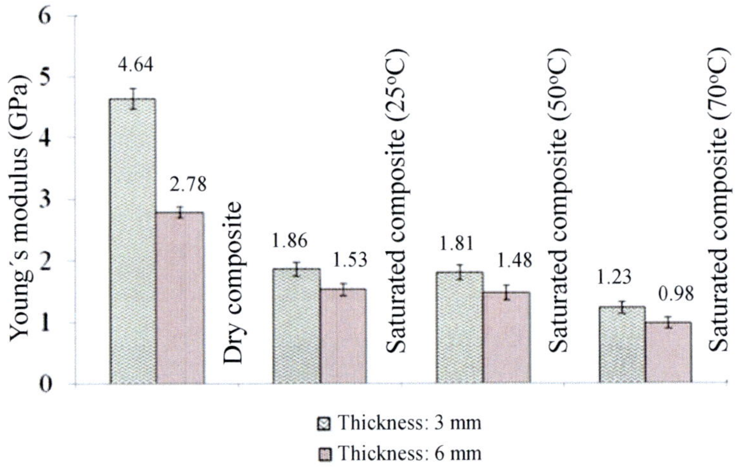

Fig. 3.19 Young's modulus of polyester composite (44.6% sisal fiber)

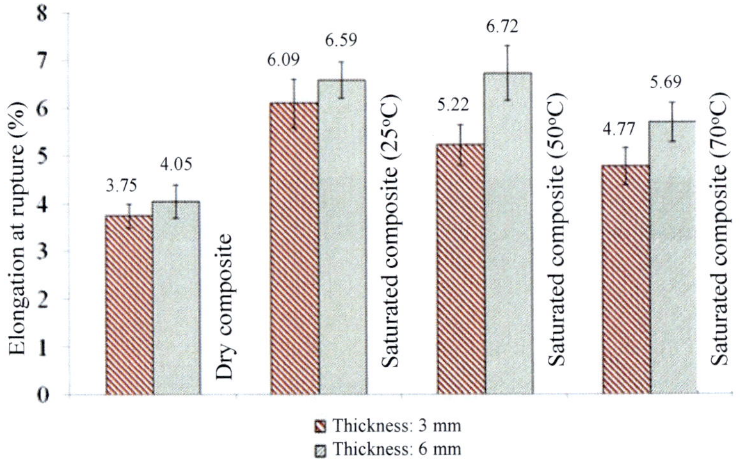

Fig. 3.20 Elongation at break (b) of polyester composite (44.6% sisal fiber)

Figures 3.21 and 3.22 illustrate the polyester-sisal samples with 3 mm and 6 mm thickness in the saturation condition at water bath temperature of 25 °C and 70 °C, respectively [20]. Tese figures show an increase in volume, as well as the deterioration of the composite samples, both due to the thermal effect and the presence of moisture. These effects are most severe at higher water bath temperature.

For composites with 3 mm thickness, there was an increase in volume of 19.77%, 21.77% and 31.34% for temperatures 25 °C, 50 °C and 70 °C, respectively. For composites with a thickness of 6 mm, the increase in volume was 15.95%, 17.25% and 21.32% for the same temperature range. The increase in volume of the samples was

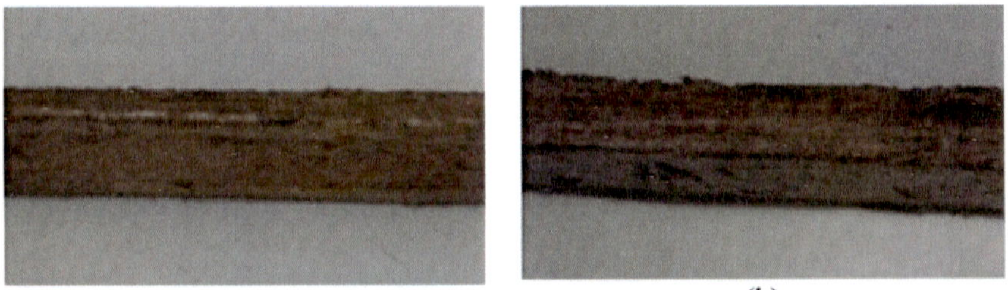

(a) (b)

Fig. 3.21 Sisal fiber-reinforced polyester composite (44.6% sisal fiber) at saturation condition (T = 25 °C). **a** 3 mm and **b** 6 mm thickness

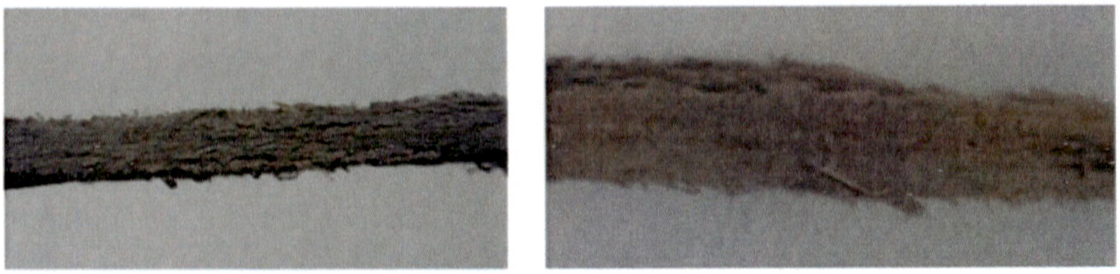

(a) (b)

Fig. 3.22 Sisal fiber-reinforced polyester composite (44.6% sisal fiber) at saturation condition (T = 70 °C). **a** 3 mm and **b** 6 mm thickness

greater than the maximum moisture content absorbed, which probably can be associated with the combined effect of the bath temperature and moisture that accelerates the sample deterioration.

With the aim to better identify the interactions between sisal fiber and polyester matrix (compatibility and dispersion), a SEM experimental study was carried out on tensile tested samples. Figures 3.23 and 3.24 illustrate the SEM micrographs of fractured surface of 3 mm and 6 mm thick untreated sisal fiber-reinforced polymer composite samples, respectively, in the saturation condition at water bath temperature 70 °C [20]. It is possible to verify that, due to the random arrangement of the short fibers and severe experimental conditions, a moderate interfacial adhesion was obtained in the composite manufactured. The main failure mechanisms in the composite samples were debonding and fiber breakage, both due to the thermal effect and the presence of moisture, as highlighted in these figures. Other phenomena such as fiber splitting, brittle surface and pull-out were also verified.

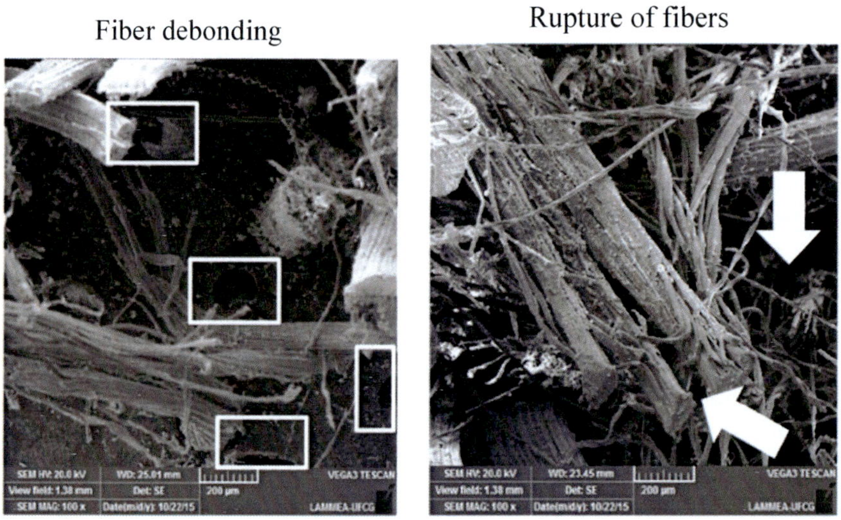

Fig. 3.23 SEM Micrographs (100X) of fractured surface of untreated sisal fiber-reinforced polyester composite (saturation condition at 70 °C and 3 mm thickness)

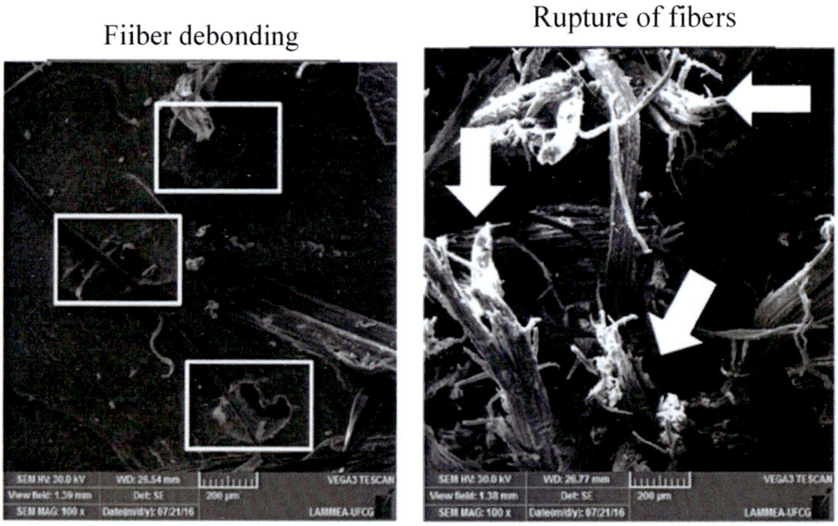

Fig. 3.24 SEM Micrographs (100X) of fractured surface of untreated sisal fiber-reinforced polyester composite (saturation condition at 70 °C and 6 mm thickness)

References

1. Herrera-Franco, P.J., Valadez-Gonzalez, A.: A study of the mechanical properties of short natural-fiber reinforced composites. Compos. Part B Eng. **36**, 597–608 (2005)
2. Chow, C.P.L., Xing, X.S., Li, R.K.Y.: Moisture absorption studies of sisal fibre reinforced polypropylene composites. Compos. Sci. Technol. **67**, 306–313 (2007)
3. Bledzki, A.K., Reihmane, S., Gassan, J.: Properties and modification methods for vegetable fibers for natural fiber composites. J. Appl. Polym. Sci. **59**, 1329–1336 (1996)
4. Clark, R.A., Ansell, M.P.: Jute and glass fibre hybrid laminates. J. Mater. Sci. **21**, 269–276 (1986)
5. Cavalcanti, W.S., Carvalho, L.H., Lima, A.G.B.: Water diffusion in unsaturated polyester composite reinforced by a hybrid jute/glass fabric: modeling and simulation. Revista Matéria **10**(1), 14–23 (2005) (In Portuguese)
6. Bledzki, A.K., Gassan, J.: Composite reinforced with cellulose based fiber. Prog. Polym. Sci. **24**, 221–274 (1999)
7. Andreopoulos, A.G., Tarantili, P.A.: Water sorption characteristics of epoxi resin–UHMPE fibers composites. J. Appl. Polym. Sci. **70**, 747–755 (1998)
8. Nóbrega, M.M.S.: Polyester matrix composite with caroá fibers (Neoglaziovia variegata): Mechanical charcterization and water sorption. Doctorate Thesis, Process Engineering, Federal University of Campina Grande, Paraíba, Brazil (2007) (In Portuguese)
9. Tsai, Y.I., Bosze, E.J., Barjasteh, E., Nutt, S.R.: Influence of hygrothermal environment on thermal and mechanical properties of carbon fiber/fiberglass hybrid composites. Compos. Sci. Technol. **69**(3–4), 432–437 (2009)
10. Marcovich, N.E., Reboredo, M.M., Aranguren, M.I.: Moisture diffusion in polyester-woodflour composites. Polym. **40**, 7313–7320 (1999)
11. Thwe, M.M., Liao, K.: Effects of environmental aging on the mechanical properties of bamboo-glass fiber reinforced polymer matrix hybrid composites. Compos. Part A Appl. Sci Manuf. **33**, 43–52 (2002)
12. Mulinari, D.R.: Thermal, mechanical and morphological behavior of high density polyethylene composites reinforced with sugarcane bagasse cellulose fiber. Thesis in Mechanical Engineering, Paulista State University (2009). Guaratinguetá, Brazil (In Portuguese)
13. Sreekala, M.S., Kumaran, M.G., Thomas, S.: Water sorption in oil palm fiber reinforced phenol formaldeyde composites. Compos. Part A Appl. Sci. Manuf. **33**, 763–777 (2002)
14. Espert, A., Vilaplana, F., Karlsson, S.: Comparison of water absorption in natural cellulosic fibres from wood and one-year crops in polypropylene composites and its influence on their mechanical properties. Compos. Part A **35**, 1267–1276 (2004)
15. Angrizani, C.A.; Vieira, C.A.B.; Zattera, A.J.; Freire, E.; Santana, R.M.C., Amico, S.C.: Influence of sisal fiber length and its chemical treatment on the properties of composites with polyester. In 17th Brazilian Congress of Engineering and Science Materials, Foz do Iguaçu, Paraná, Brazil (2006) (In Portuguese)
16. Fernandes, R.I.M.: Development and characterization of polypropylene/glass fiber/pineapple fiber composites. Ph.D Thesis in Mechanical Engineering, Paulista State University (2010). Guaratinguetá, Brazil (In Portuguese)
17. Vieira, C.A.B.: Evaluation of manufacturing methods of hybrid fabrics of short glass and sisal fibers in polymer composites, Master Dissertation in Engineering and Materials Science, University of Caxias do Sul (2008). Caxias do Sul, Brazil (In Portuguese)
18. Jayamol, G., Bhagawan, S.S., Thomas, S.: Effects of environment on the properties of low-density polyethylene composites reinforced with pineapple-leaf fibre. Compos. Sci. Technol. **58**, 1471–1485 (1998)
19. Cavalcanti, W.S.; Carvalho, L.H.; Alsina, O.L.S.; Lima, A.G.B.: Water sorption from insulated polyester composite reinforced with jute and jute/glass: modeling, simulation and experimentation. Polym. Sci. Technol. **20**, 78–83 (In Portuguese)

20. Santos, D.G.: Thermo-hydric study and mechanical characterization of polymeric matrix composites reinforced with vegetable fiber: 3D simulation and experimentation. Doctoral Thesis in Process Engineering, Federal University of Campina Grande (2017). Campina Grande, Brazil
21. Sanchez, E.M.S., Calvani, C.S., Leal, C.V., Sanchez, C.G.: Unsaturated polyester resin composite reinforced with sugarcane bagasse: influence of the treatment of fibers on properties in the composite. Polym. Sci. Technol. **20**, 194–200 (In Portuguese)
22. Tita, S.P.S., Paiva, J.M.F., Frollini, E.: Impact strength and other properties of lignocellulosic composites: phenolic thermoset matrices reinforced with sugarcane bagasse fibers. Polym. Sci. Technol. **12**:228–239 (In Portuguese)
23. Rao, R.M.V.G.K., Balasubramanian, N., Chanda, M.: Moisture absorption phenomenon in permeable fiber polymer composites. J. Appl. Polym. Sci. **26**, 4069–4079 (1981)
24. Semsarzadeh, M.A., Amiri, D.: Binder for jute-reinforced unsaturated polyester resin. Polym. Eng. Sci. **25**, 618–619 (1985)
25. Idriss Ali, K.M., Uddin, M.K., Bhutyan, M.I.U., Khan, M.A.: Improvement of jute fiber through ultraviolet-cured film of uretane acrylate. J. Appl. Polym. Sci. **54**, 303–308 (1994)
26. Soni, R.P., Soni, M.: Studies on natural fibre reinforced phenolic composites. J. Sci. Ind. Res. **58**, 34–36 (1999)
27. Srihari, S., Revathi, A., Rao, R.M.V.G.K.: Hygrothermal effects on RT-Cured glass-Epoxi composites in immersion environments. Part A: moisture absorption characteritics. J. Reinf. Plast. Compos. **21**, 983–991
28. Pavlidou, S., Papaspyrides, C.D.: The effect of hygrothermal history on water sorption and interlaminar shear strength of glass/polyester composites with different interfacial strength. Compos. Part A Appl. Sci. Manuf. **34**, 1117–1124 (2003)
29. Pegoretti, A., Penati, A.: Effects of hygrothermal aging on the molar mass and thermal properties of recycled Poly(Ethylene Terephathalate) and its short glass fibre composites. Polym. Degrad. Stab. **86**, 233–243 (2004)
30. Kumosa, L., Benedikt, B., Armentrout, D., Kumosa, M.: Moisture absorption properties of unidirectional glass/polymer composites used in composite (non-ceramic) insulators. Compos. Part A **35**, 1049–1063 (2004)
31. Nóbrega, M.M.S., Cavalcanti, W.S., Carvalho, L.H., Lima, A.G.B.: Water absorption in unsaturated polyester composites reinforced with caroá fiber fabrics: modeling and simulation. Mat.-wiss. u.Werkstofftech. **41**(5), 300–305 (2010)
32. Cruz, V.C.A., Nóbrega, M.M.S., Silva, W.P., Carvalho, L.H., Lima, A.G.B.: An experimental study of water absorption in polyester composites reinforced with macambira natural fiber. Mat.-wiss. u.Werkstofftech. **42**(11), 979–984 (2011)
33. Pimentel, J.R.M.: Characterization and properties analysis of the macambira fiber (Bromelia laciniosa). Master Dissertation in Mechanical Engineering, Federal University of Rio Grande do Norte (2012). Natal, Brazil
34. Silva, C.J.: Water absorption in composite materials of vegetal fiber: modeling and simulation via CFX. Master Dissertation in Mechanical Engineering, Federal University of Campina Grande (2014). Campina Grande, Brazil
35. Joseph, P.V., Rabello, M.S., Mattoso, L.H.C., Joseph, K., Thomas, S.: Environmental effects on the degradation behaviour of sisal fibre reinforced polypropylene composites Compos. Sci. Technol. **62**, 1357–1372 (2002)
36. Cavalcanti, W.S.: Polyester/hybrid jute-glass fabric composites: Mechanical characterization and water sorption simulation Doctorate Thesis. Federal University of Campina Grande, Paraíba, Brazil, Process Engineering (2006).(In Portuguese)

37. Santos, D.G., Lima, A.G.B., Pinto, M.V.: Silva mechanical characterization and water sorption in polyester matrix composites reinforced with sisal fiber: an experimental investigation. Defect Diffus. Forum. **369**, 131–134 (2016)
38. Du, Y., Yan, N., Kortschot, M.T.: The use of ramie fibers as reinforcements in composites. In: Faruk , O., Sain, M (Eds.) Biofiber Reinforcement in Composites Materials, Elsevier Ltda., Amsterdam, Chapter 4, pp. 104–137 (2015)
39. Carvalho, L.H., Moraes, G.S., d'Almeida, J.R.M.: Influence of water absorption and pre-drying conditions on the tensile mechanical properties of hybrid lignocellulosic fiber/polyester composites. J. Reinf. Plast. Compos. **28**(16), 1921–1932 (2009)

Chapter 4
Fick's Model Analysis

In this chapter, based on the rigorous theory of pure diffusion through fibrous porous media (Fick's second law), different approaches (analytical, numerical and by CFD) and its limitations for the correct description of water absorption process in vegetable fiber-reinforced polymer composites will be presented. In the macroscopic mathematical modeling both fiber and polymer are considered to be a homogeneous mixture and the effect of swelling was not taken into account. Herein, by using a three-dimensional mathematical treatment, different effects of some process parameters (sample thickness and water bath temperature) in the moisture migration behavior inside the material were analyzed. Applications were perfomed on polyester composites reinforced by caroá and macambira vegetable fibers.

4.1 Transport Phenomena in Porous Media: Foundations

Transport phenomena in porous media represent an important research area related to heat and mass transfer, and fluid flow fields. Heat and mass diffusion/convection through any heterogeneous media, depends on the structure of the matrix and the physical properties of each phase (fluid and solid) involved in the physical problem.

For this class of problem, one of the most difficult aspects of the analysis is structural modeling of the domain. A theoretical treatment based upon the traditional method starts with a potential per unit time balance in a differential control volume with arbitrary shape in a macroscopic scale (i.e. porous material is represented as a fictitious continuum).

By assuming local equilibrium between solid and fluid, the heat and mass transfer phenomena in porous media can be modeled by the macroscopic conservation equation in a short form, as follows:

$$\frac{\partial}{\partial t}(\lambda\Phi) + \nabla.(\lambda\vec{v}\Phi) = \nabla.(\Gamma^{\Phi}\nabla\Phi) + S^{\Phi} \tag{4.1}$$

J. M. P. Q. Delgado et al., *Moisture Transport in Polymer Composite Materials*, Advanced Structured Materials 160, https://doi.org/10.1007/978-3-030-77826-2_4

For mass transfer problems, in the Eq. (4.1), we have $\lambda = \rho$; $\Phi = M$ and $\Gamma^{\Phi} = \rho D$, where ρ, M, and D correspond to density, moisture content, and effective mass diffusivity, respectively, and t is the time. In this same equation, $\vec{v}$ is the velocity vector and S^{Φ} is a mass source term.

Mass diffusion is a phenomenon in which matter is transported from one region in space to another due to random molecular motions [1]. Heat and mass transfer creates temperature and moisture gradients inside the solid, which in general, depends on the internal and external conditions of the porous media. For simplicity, several researchers report a model that assumes that water migrates at very low velocity from the external surrounding medium to the center of the solid only by liquid diffusion (diffusive effects are greater than convective effects). Further, gravity, capillarity, and other effects are neglected.

In practical applications, the diffusion process of water in polymeric composites occurs slowly. Generally, the time until moisture saturation is reached in the composite is a few months. The process depends on fiber content, size and shape of the composite as well as on matrix identity and environmental conditions, especially temperature, as already mentioned. In the case of water vapor absorption from the air, relative humidity also plays an important role. Thus, in order to obtain faster information and with lower process costs, researchers have studied the water absorption process using numerical simulation from developed mathematical models. The main one reported in the literature is the well known as Fick's model.

4.2 Moisture Absorption by Fick's Model

4.2.1 The Physical Problem

For the analysis of the physical problem studied here, consider a dry fiber-reinforced polymer composite of parallelepiped geometry, at a given temperature, which is suddenly immersed in a stationary fluid (water) at a different and higher temperature. As the heated fluid surrounding the dry porous media heats it, heat penetrates into the solid (as the result of a temperature difference) and moisture migrates into the solid by diffusion from the surface (due to a concentration difference). Figure 4.1 sechematically illustrates the problem treated here.

4.2.2 The General Mass Diffusion Equation

In this study, it is assumed that moisture sorption occurs only by diffusion. According to Fick's first law, the diffusive flux (J) is directly proportional to the concentration gradient. Thus, the following equation can be written:

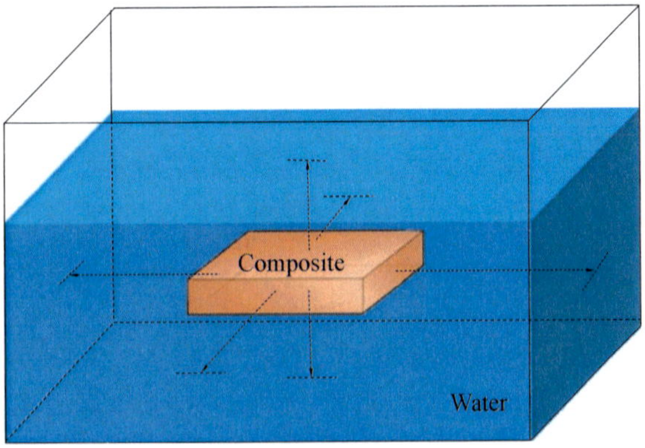

Fig. 4.1 Squematic of a composite sample immersed in a stationary fluid

$$J = -D\frac{\partial M}{\partial y} \tag{4.2}$$

In situations where the solid density may be considered constant and without mass generation inside the solid, the partial differential equation that describes the mass diffusion phenomenon (for example, water absorption), well known as Fick's second law of diffusion, can be written as follows:

$$\frac{\partial M}{\partial t} = \nabla \bullet (D\nabla M) \tag{4.3}$$

where D is the effective mass diffusion coefficient and M is the moisture content in a dry basis (mass of absorbed water per mass of the dry solid).

Based in the Fick's model, several works related to moisture absorption in composite systems have presented analytical and/or numerical solutions of the governing equation. However, all these works consider the diffusion through the solid to be one-dimensional [2–9]. Yao and Ziegmann [7] related the water absorption behavior of three types of glass-fiber reinforced plastic (GRP) pipes. A model was developed from Fick's law to predict water diffusion curves of the GRP pipe specimens. It was found that only one of the GPR pipe specimens exhibited near-Fickian behavior and the other two behaved like a two-stage diffusion process because of the abnormal non-Fickian process. Srihari et al. [6] concluded in their studies that the composite (glass/epoxy) and neat resin casting specimens immersed in distilled water and artificial seawater exhibited Fickian behavior. Najafi et al. [8] studied water absorption of wood plastic composites. They concluded that water absorption in these composites follows a Fickian diffusion process. Czél and Czigány [10] in their study of water absorption of glass fiber/polyester composite pipes concluded that the Fickian diffusion model did not fit well the absorption curves, but a special explicit, asymptotically correct function based on the Lucas-Washburn equation was found to be more adequated to describe water uptake process.

In some cases, in order to take into consideration three-dimensional problems, some researchers have used analytical solutions applicable for long times which greatly simplify the problem [11–13]. Chateauminois et al. [11] studied the water diffusion in solids with parallelepiped shape using the short term solution of the one-dimensional Fick's law. For longer times, the water absorption kinetics was computed using the three-dimensional solution of Fick's law. Pavan et al. [12] in their studies on moisture diffusion behavior of glass/epoxy composites found that the diffusion parameters obtained from the experiment conducted up to saturation limits agreed well with the data computed. Bao and Yee [13] showed in their studies of moisture absorption based on longterm absorption data that the fast Fickian diffusion is followed by a slow gradual increase in weight gain. This diffusion behavior can be successfully described by the proposed two-stage diffusion model. In this two-stage model, the first and second stage of diffusion are assumed to be diffusion and relaxation-controlled, respectively.

For a more general formulation, Cavalcanti et al. [14] and Cavalcanti [15] report a three-dimensional mathematical modeling approach to describe moisture absorption in an unsaturated polyester composite reinforced by a hybrid jute/glass fabrics. Predicted results on the average moisture constant inside the material during the whole process were compared with experimental data, in order to validate the model and to obtain the effective diffusion coefficient. According to the authors a good agreement was obtained in all cases evaluated. Also, in complement of those researches, the works of Santos [16], Santos et al. [17], Melo [18], and Carvalho et al. [19] can be cited.

4.2.3 The Mass Diffusion Equation: 3D Approach in Cartesian Coordinates

Based on the hypotheses already reported, the mass transfer in a solid of parallelepiped shape of dimensions $2R_1 \times 2R_2 \times 2R_3$ (Fig. 4.2), the mass conservation equation written in Cartesian coordinates for the three-dimensional case, is given as:

$$\frac{\partial M}{\partial t} = \frac{\partial}{\partial x}\left(D\frac{\partial M}{\partial x}\right) + \frac{\partial}{\partial y}\left(D\frac{\partial M}{\partial y}\right) + \frac{\partial}{\partial z}\left(D\frac{\partial M}{\partial z}\right) \tag{4.4}$$

For a well-posed mathematical model it is necessary to specify the initial and boundary conditions. Because the symmetry of the problem, we will consider only 1/8 of the volume of the solid. In this chapter, the initial, symmetry and boundary conditions are as follows:

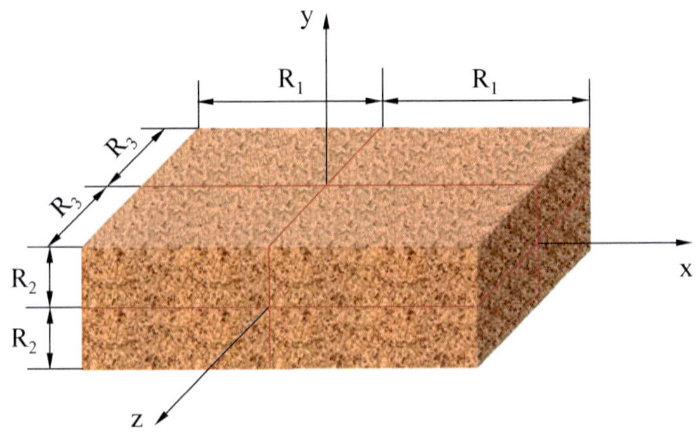

Fig. 4.2 Geometrical configuration of the physical domain

(a) Initial condition:

$$M(x, y, z, t = 0) = M_o, \quad \begin{cases} 0 < x < R_1 \\ 0 < y < R_2 \\ 0 < z < R_3 \end{cases} \tag{4.5}$$

(b) Symmetry conditions:

$$\frac{\partial M(x = 0, y, z, t)}{\partial x} = \frac{\partial M(x, y = 0, z, t)}{\partial y} = \frac{\partial M(x, y, z = 0, t)}{\partial x} = 0, t > 0 \tag{4.6}$$

(c) Boundary conditions:

$$M(x = R_1, y, z, t) = M(x, y = R_2, z, t) = M(x, y, z = R_3, t) = M_e, t > 0 \tag{4.7}$$

The average moisture content of the solid at any instant of the process is computed as:

$$\overline{M}(t) = \frac{1}{V} \int_V M(x, y, z, t)\, dV \tag{4.8}$$

where V is the volume of the porous solid.

In general, the diffusion coefficient is considered constant or dependent the local moisture content, average moisture content or temperature, or yet as a combination between of moisture content and temperature. Then, based on these assumptions, the following equations can be used for this purpose:

$$\mathrm{D} = \mathrm{D_o} \tag{4.9}$$

$$\mathrm{D} = \mathrm{D_1}\,Exp(\mathrm{c_1 M}) \tag{4.10}$$

$$\mathrm{D} = \mathrm{D_2 Exp}\left(\mathrm{c_2}\overline{\mathrm{M}}\right) \tag{4.11}$$

$$\overline{\mathrm{D}} = \frac{1}{(\mathrm{M_e} - \mathrm{M_o}}\int_{\mathrm{M_o}}^{\mathrm{M}} \mathrm{D}(\overline{\mathrm{M}})\mathrm{d}\overline{\mathrm{M}} \tag{4.12}$$

$$\mathrm{D} = \mathrm{D_3 Exp}\left(\frac{\mathrm{c_3}}{\mathrm{T}}\right) \tag{4.13}$$

where $\mathrm{D_i}$ and $\mathrm{c_i}$ are constants, obtained by fitting with experimental data.

4.2.4 Solution Techniques: Three-Dimensional Approach

To solve the diffusion partial differential equation (Eq. 4.4) in conjunction with appropriate initial and boundary conditions, different analytical (for example, separation of variables and Galerkin-based integral method) and numerical (for example, finite-difference, finite-element, boundary-element, and finite-volume) techniques may be used. The choice of a particular technique depends on the easier procedure for a particular physical situation.

4.2.4.1 Analytical Solution

Based on a review of the technical literature, it can be verified that some works analytically solve the diffusion problems with prescribed boundary conditions (M = Me) and constant thermo-physical properties (D = cte). For this particular situation and using separation of variables technique, the analytical solution of the mass diffusion equation applied to a parallelepiped is given as [20, 21]:

$$\frac{M(x,y,z,t) - M_e}{M_o - M_e} = \sum_{n=1}^{\infty}\sum_{m=1}^{\infty}\sum_{k=1}^{\infty} A_n A_m A_k \times \mathrm{Cos}(\beta_n x) \times \mathrm{Cos}(\beta_m y) \times \times \mathrm{Cos}(\beta_k z) \times Exp\left[-\left(\beta_n^2 + \beta_m^2 + \beta_k^2\right)Dt\right] \tag{4.14}$$

where

$$A_n = \frac{2Sen(\beta_n R_1)}{\beta_n R_1} \tag{4.15}$$

$$A_m = \frac{2Sen(\beta_m R_2)}{\beta_m R_2} \tag{4.16}$$

$$A_k = \frac{2Sen(\beta_k R_3)}{\beta_k R_3} \tag{4.17}$$

The moisture content distribution inside the porous media at different times can be obtained using Eq. (4.14). In this equation, the coefficients β_n, βm and β_k are the so-called eigenvalues. They are obtained using the following equations devived from the boundary conditions previously established:

$$Cos(\beta_n R_1) = 0 \tag{4.18}$$

$$Cos(\beta_m R_2) = 0 \tag{4.19}$$

$$Cos(\beta_k R_3) = 0 \tag{4.20}$$

According to Eq. (4.8), the average moisture content can then be calculated as:

$$\frac{\overline{M}(t) - M_e}{M_o - M_e} = \sum_{n=1}^{\infty}\sum_{m=1}^{\infty}\sum_{k=1}^{\infty} B_n B_m B_k . e^{-[\beta_n^2 + \beta_m^2 + \beta_k^2]Dt} \tag{4.21}$$

where

$$B_n = \frac{2}{(\beta_n R_1)^2} \tag{4.22}$$

$$B_m = \frac{2}{(\beta_m R_2)^2} \tag{4.23}$$

$$B_k = \frac{2}{(\beta_k R_3)^2} \tag{4.24}$$

4.2.4.2 Numerical Solution

It may be difficult to obtain the analytical solution of a partial differential equation with a high level of complexity as, for example, when variable thermo-physical properties are used. In this case a numerical solution may be used as an appropriate alternative technique. For this purpose, the finite-volume method has been frequently used to discretize the governing equations. In this tecnique, the physical domain is transformed in a discrete domain formed by a finite number of control-volumes. Figure 4.3 shows a structured grid that represents the physical domain in a discretized form. This is only a representative squematics.

Figure 4.4 illustrates the differential volume of the physical and computational domains, Figs. 4.2 and 4.3, respectively, where the nodal points (W, E, N, S, F, T), the dimensions of the control volume (Δx, Δy, Δz) and the distances between nodal points (δx, δy, δz) are presented.

Assuming a fully implicit formulation, where all terms are estimated in $t + \Delta t$, Eq. (4.4) was integrated in the control volume of Fig. 4.4, which corresponds to the

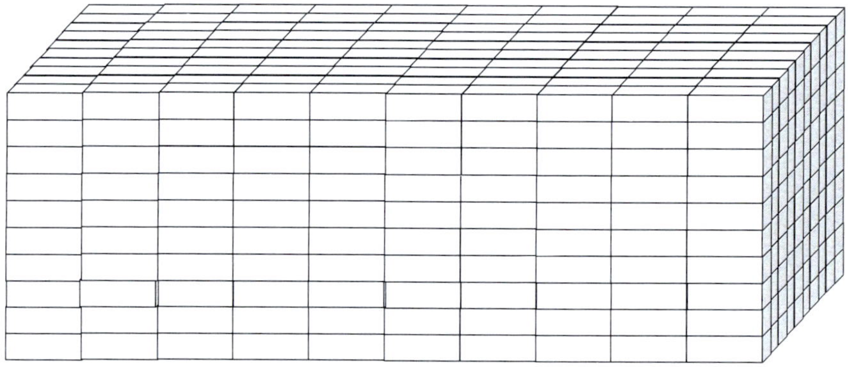

Fig. 4.3 The schematic illustration of a structured numerical grid

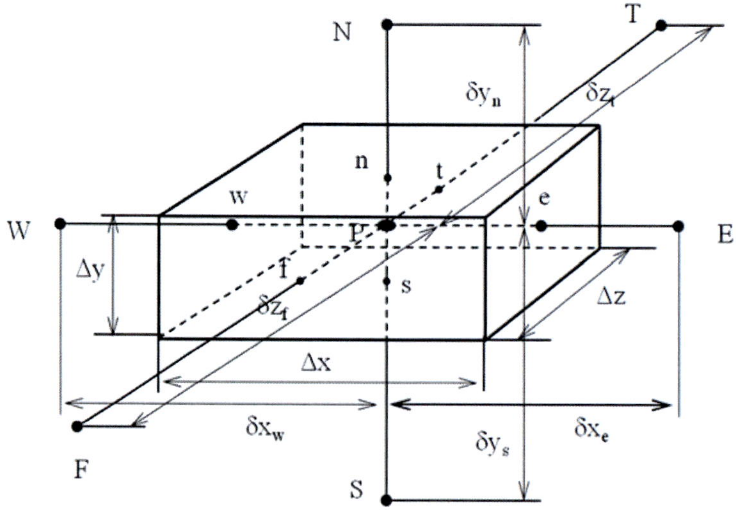

Fig. 4.4 Control volume used in this work

internal points of the domain, and also, in time. As a result, Eq. (4.4) was discretized by using practice B (nodal points in the center of control-volume) [22–24] resulting in a linear algebraic equation as follows:

$$A_P M_P^* = A_E M_E^* + A_W M_W^* + A_N M_N^* + A_S M_S^* + A_T M_T^* + A_F M_F^* + B \quad (4.25)$$

where

$$M^* = \frac{M - M_e}{M_o - M_e} \quad (4.26)$$

$$A_E = \frac{D_e \Delta y \Delta z}{\delta x_e} \quad (4.27)$$

$$A_W = \frac{D_w \Delta y \Delta z}{\delta x_w} \quad (4.28)$$

$$A_N = \frac{D_n \Delta z \Delta x}{\delta y_n} \quad (4.29)$$

$$A_S = \frac{D_s \Delta z \Delta x}{\delta y_s} \quad (4.30)$$

$$A_T = \frac{D_t \Delta x \Delta y}{\delta z_t} \quad (4.31)$$

$$A_F = \frac{D_f \Delta x \Delta y}{\delta z_f} \quad (4.32)$$

$$A_P^o = \frac{\Delta x \Delta y \Delta z}{\Delta t} \quad (4.33)$$

$$B = A_P^o M_p^{*o} \quad (4.34)$$

$$A_P = A_E + A_W + A_N + A_S + A_T + A_F + A_P^o \quad (4.35)$$

In the discretized form, the average moisture contents can be written, as follows:

$$\overline{M} = \frac{1}{V} \sum_{i=2}^{npx-1} \sum_{j=2}^{npy-1} \sum_{k=2}^{npz-1} M_{i,j,k} \Delta V_{i,j,k} \quad (4.36)$$

in which i, j e k represent the position of the nodal point in the x, y, and z directions, respectively, and *npx*, *npy* and *npz* are the nodal point numbers in the x, y, and z directions, respectively.

According to the literature [22], the mass diffusion coefficient at the interface of the control-volume is calculated as:

$$D_i = \frac{2D_P D_E}{D_P + D_E} \quad (4.37)$$

where the subscript i refers to the interface, and D_P and D_E are the values of the mass diffusion coefficient in the nodal points P and E, respectively.

The set of algebraic linear equations obtained from Eq. (4.25) (when applied for all internal nodal points) can be solved interactively, for example, using the Gauss–Seidel method. The following convergence criterion is frequently used:

$$\left|M^{*n-1} - M^{*n}\right| \leq 10^{-8} \quad (4.38)$$

where n represents the n-th iteration in each time.

In practical applications, because of the large variety of materials and operating conditions, in general, the diffusion coefficient is unknown. Thus, in general, this process parameter is found by varying it to minimize the sum of the squared deviations between the actual and predicted data (the least square error technique). The relative deviation between the experimental and calculated values (relative residual, ERMQ) and the variance (S^2) are defined as [25]:

$$ERMQ = \sum_{i=1}^{m}\left(\overline{M}^*_{i,\,Numerical} - \overline{M}^*_{i,\,Numerical}\right)^2 \quad (4.39)$$

$$S^2 = \frac{ERMQ}{(m-1)} \quad (4.40)$$

where m is the number of experimental points. The smallest values of ERMQ and S^2 are used as criteria to obtain the best value of the diffusion coefficient for each operating condition. This numerical formulation was used to predict the water uptake in caroá and macambira fiber-reinforced polyester composite; the operating conditions and experimental data are reported in Chap. 3, in details.

4.2.4.3 Solution by Commercial Software (Ansys CFX)

In this topic, the software Ansys CFX was used to simulate the water absorption process in caroá fiber-reinforced polyester composites. The study was developed and reported in detail in Silva [26, 27] and Silva et al. [28]. The geometry considered in this research is of a solid parallelepiped with dimensions as illustrated in Fig. 4.2.

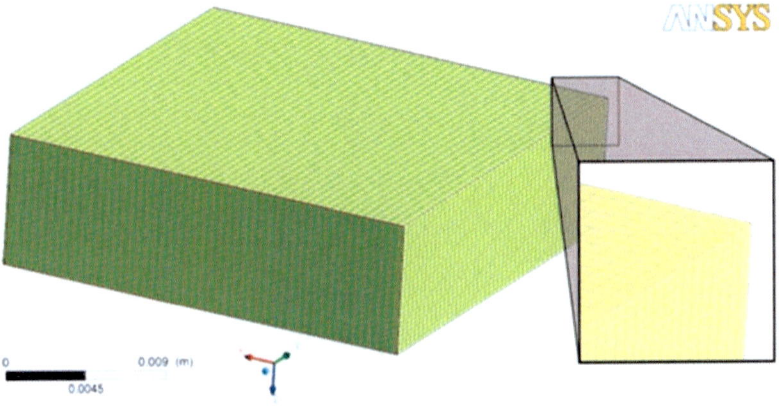

Fig. 4.5 Numerical mesh used in this study

In this work, the mesh built using the ICEM CFD (ANSYS CFX®) software ($6 \times 20 \times 20\ mm^3$), is composed of 75,000 nodal points and 60,623 hexahedral elements. Figure 4.5 illustrates the geometry, the numerical grid and detail of regions close to the surface of the composites. For the transient numerical solution, a time step of 200 s and a 10^{-5} convergence criterion was used. The operating conditions and experimental data were reported in Chap. 3.

4.2.5 *Fick's Model Application*

4.2.5.1 Analytical Results

(a) Water Sorption Kinetics

Santos [16] and Santos et al. [17] reported an analytical study about water absorption in sisal fiber-reinforced polyester composites using Eq. (4.21). The analytical results of the average moisture content of the composites were compared with the experimental data at different temperatures and sample thicknesses. Figures 4.6, 4.7 and 4.8 illustrate these comparisons. For this comparison, the authors used 30 terms of the infinite series reported in Eq. (4.21). Upon analyzing these figures we can see a good agreement between the results. Variations between the experimental and analytical data can be attributed to several factors such as: the nature of the studied composites, i.e. vegetable fiber has hygroscopic nature, the lack of uniformity in the diameter of sisal fibers, and the small temperature variations during measurements, even if the tests were carried out at controlled bath temperatures. Furthermore, it can be noted that after the samples were immersed in water, the moisture absorption occurs quite rapidly in the first 50 h. These effects were more pronounced at higher water bath temperature.

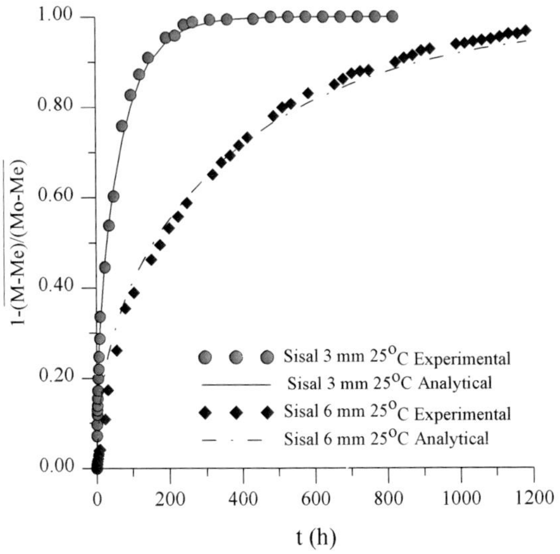

Fig. 4.6 Dimensionless average moisture content during the water absorption in sisal fiber reinforced unsaturated polyester composites, at 25 °C

Fig. 4.7 Dimensionless average moisture content during the water absorption in sisal fiber reinforced unsaturated polyester composites, at 50 °C

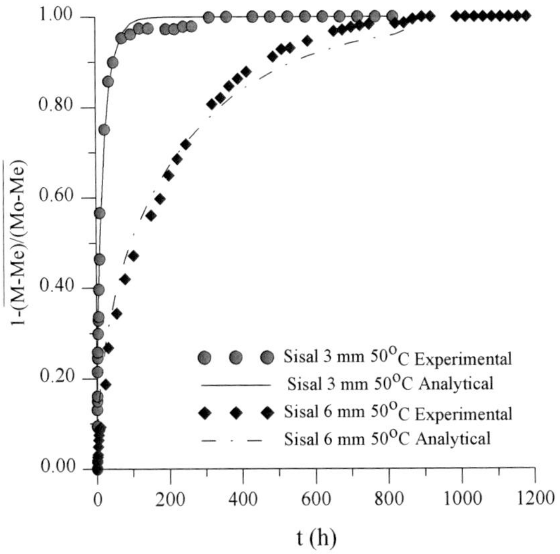

For longer times, the curve gradually changes its slope until levelling off, reaching moisture saturation level. This relief in the absorption rate is attributed to how water molecules moves into the fibers by capillarity which is in its surface or into the lumen. In 6 mm thick samples, maximum saturation times were longer, which may be associated to the way the moisture is transmitted between fibers. Initially moisture

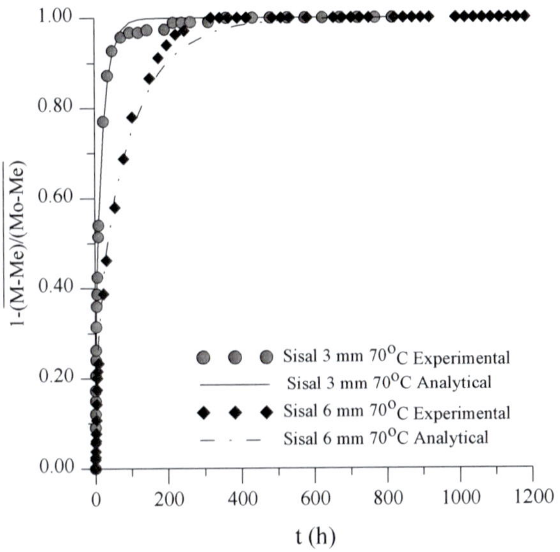

Fig. 4.8 Dimensionless average moisture content during the water absorption in sisal fiber reinforced unsaturated polyester composites, at 70 °C

is absorbed on the fiber surface and then migrates to the lumen where it proceeds by capillary action so that the diffusive process is maintained. Since thicker samples have higher volume, the maximum saturation time is also higher, which justifies the data obtained. Furthermore, it was verified that, for the thinner composites (3 mm thick), water absorption is faster, which is though to be due to the greater area/volume ratio displayed by these samples.

This phenomenon can be attributed to the thermal and geometric expansions of the matrix (variation in the area/volume ratio). As the water bath temperature increases, so does the kinetic energy of the water molecules that, in turn cause, a decrease in water intermolecular forces decreasing its viscosity and thus favoring its migration into the composite.

According to those authors [16, 17], the effect of moisture and temperature has been studied using as reference different parameters such as: tensile and shear strengths, elastic modulus, fatigue behavior, creep, stress rupture, response to dynamic impact, electrical resistance, and dimensions variations (swelling). These studies indicated that composite moisture content and water bath temperature up to about glass-transition temperature have provoked small effects on fiber-dominated composite properties, but significantly reduce matrix-dominated composite properties.

4.2.5.2 Numerical Results

(a) Absorbed Water Kinetics

Numerical results of average moisture content of the composites under investigation were compared with the experimental data as reported in Chap. 3. To obtain the numerical results, a computational code using 20 × 20 × 20 nodal points and a time step of $\Delta t = 1$ s was developed. These parameters were obtained after a time step and mesh refining study [15]. The numerical results obtained from a given mathematical model are strongly dependent on the boundary conditions, thermo-physical properties and geometry. A comparison between the experimental and numerical predicted average moisture content for the investigated caroá/polyester composites, shown in Figs. 4.9, 4.10 and 4.11, and macambira/polyester composites, shown in Figs. 4.12, 4.13, indicates that the model employed here is appropriate. These figures, show that a good agreement was obtained in all cases investigated.

In general, the curves are similar to those of other natural fiber reinforced polymer matrix composites. For all investigated composites the moisture content increases monotonically with water immersion time until it reaches the equilibrium moisture content (saturation condition).

A comparison between Figs. 4.9, 4.10, 4.11, 4.12 and 4.13 indicates that the wetting process in the 3-mm thick composite was faster than in the 6-mm thick composite. This behavior may be attributed to the increase in area/volume ratio and the high water affinity for the cellulosic material. The effect of the thickness sample is more pronounced at lower temperatures. At high temperature the kinetics of water absorption is dominated by thermal effects, and moisture diffusion inside the material is strongly dependent on the temperature.

Fig. 4.9 Dimensionless average moisture content during the water absorption in caroá fiber reinforced unsaturated polyester composites, at 25 °C

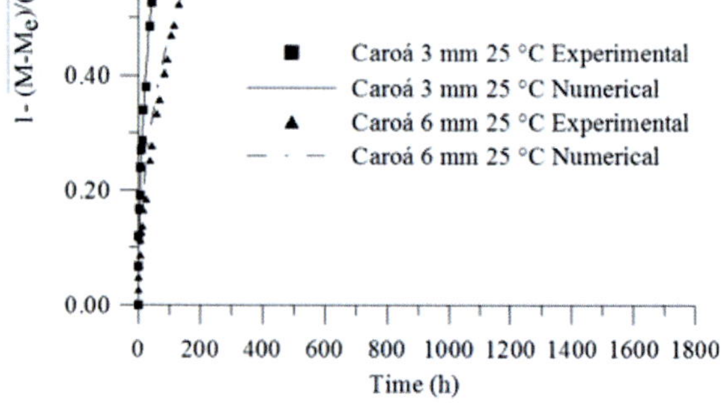

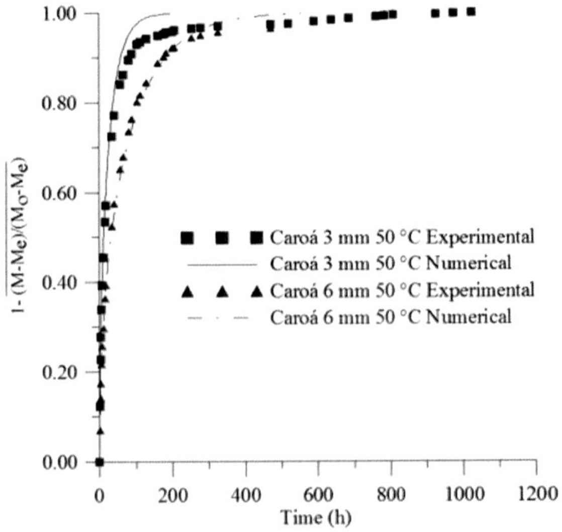

Fig. 4.10 Dimensionless mean moisture content during the water absorption in caroá fiber reinforced unsaturated polyester composites, at 50 °C

Furthermore, the results indicate that, the water uptake of the composite sample immersed in a water bath at 70 °C was faster than under the other (25 and 50 °C) experimental conditions adopted. This behavior is attributed to the increased water mobility within the solid at higher temperature due to thermal dilation and increased composite porosity (thermo-activation).

According to the data reported in the literature, it was verified that water absorption for the composites is higher than for the matrix. Neat polyester shows a very low

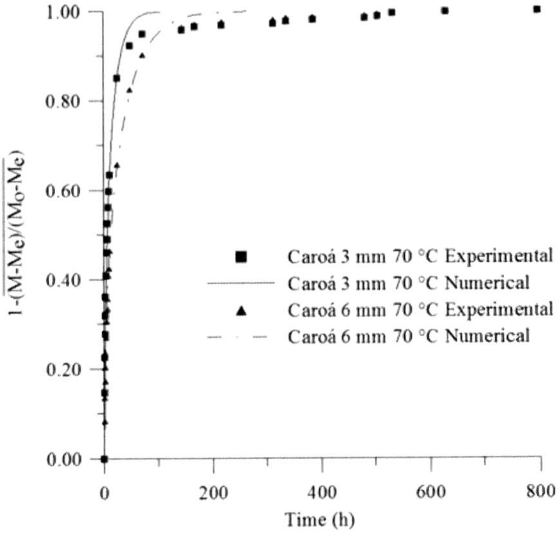

Fig. 4.11 Dimensionless mean moisture content during the water absorption in caroá fiber reinforced unsaturated polyester composites, at 70 °C

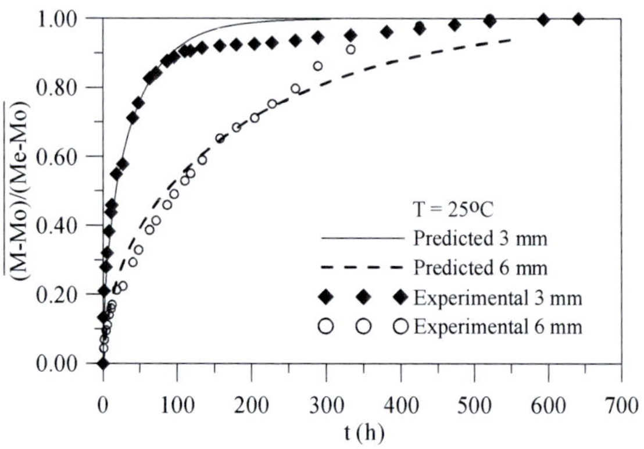

Fig. 4.12 Dimensionless average moisture content during the water absorption in macambira fiber reinforced unsaturated polyester composites, at 25 °C

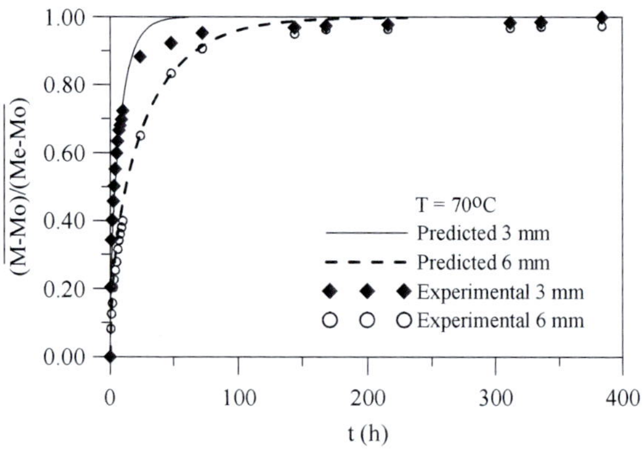

Fig. 4.13 Dimensionless average moisture content during the water absorption in macambira fiber reinforced unsaturated polyester composites, at 70 °C

percentage of absorbed water. While the water sorption in equilibrium for the unsaturated polyester is about 1.2%, for the composites it was about 14%. Degradation or delamination was not verified in the composites after each experiment.

These results are consistent with those reported in similar systems for unsaturated polyester/jute composites [14, 15] unsaturated polypropylene/sisal composites [29, 30], and cabonfiber/fiberglass hybrid composites [31].

It is important to notice that moisture degrades natural fiber reinforced composites, but this effect was not measured here. However, weight loss varying with immersion time and fiber content was verified for dried sisal fiber reinforced polypropylene composites [30].

(b) Absorbed Water Distribution

The dimensionless moisture content distribution, $(M - M_e)/(M_o - M_e)$, in the 70% unsaturated polyester/30% caroá fiber composites in the plane x = 5 mm, at elapsed time of 20 h, is presented in Figs. 4.14 and 4.15, for 3 and 6 mm thickness, respectively. Figures 4.16 and 4.17 illustrate the same results for unsaturated polyester composites reinforced with macambira natural fiber. Upon analyzing these

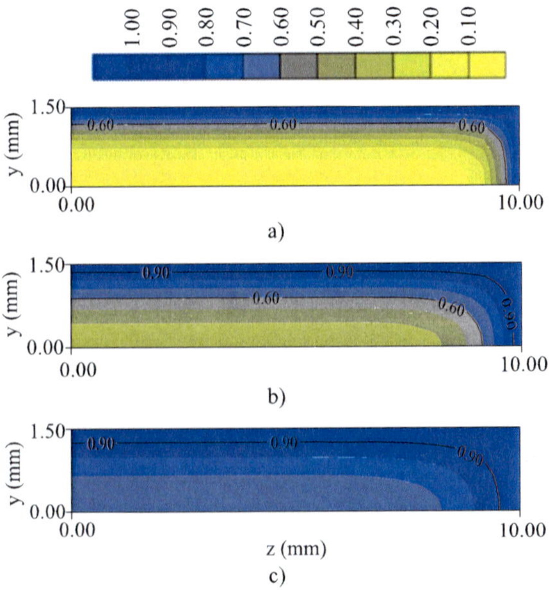

Fig. 4.14 Moisture content distribution inside the composite 30% caroá fiber/70% unsaturated polyester, at 25 °C (**a**), 50 °C (**b**) and 70 °C (**c**) for a thickness of 3 mm, in the plane x = 5 mm, elapsed time 20 h

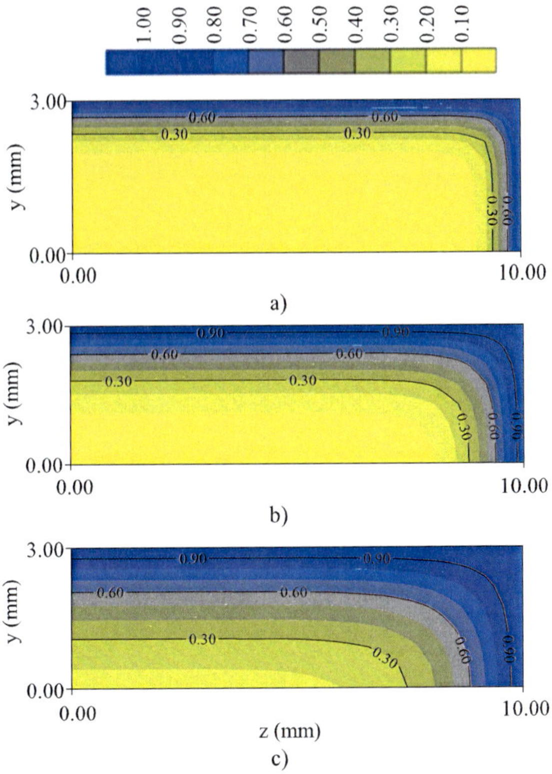

Fig. 4.15 Moisture content distribution inside the composite 30% caroá fiber /70% unsaturated polyester, at 25 °C (**a**), 50 °C (**b**) and 70 °C (**c**) for a thickness of 6 mm, in the plane x = 5 mm, elapsed time 20 h

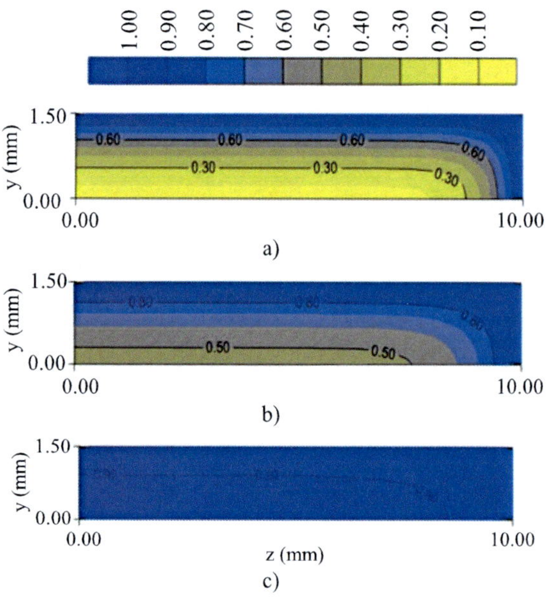

Fig. 4.16 Dimensionless moisture content distribution inside the macambira composite (3 mm thickness, x = 5 mm, elapsed time 20 h). **a** 25 °C, **b** 50 °C and **c** 70 °C

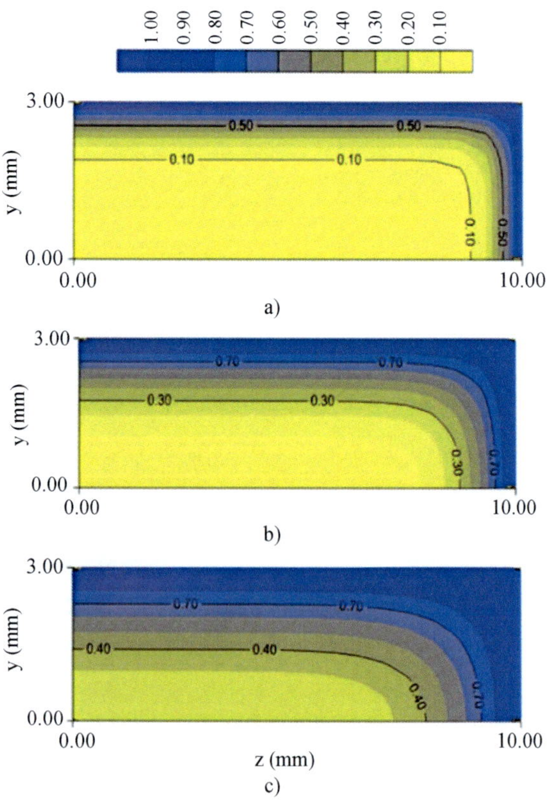

Fig. 4.17 Dimensionless moisture content distribution inside the macambira composite (6 mm thickness, x = 5 mm, elapsed time 20 h). **a** 25 °C, **b** 50 °C and **c** 70 °C

figures, we can see the existence of high moisture gradients closed to the vertices of the composite. These figures also show the influence of temperature on the water absorption rate. It can be observed that 20 h after the test had started, the moisture gradients at higher temperature is significantly higher for both 3 and 6 mm thick composites studied.

Figure 4.18 illustrates the dimensionless moisture content distribution inside the macambira fibe-reinforced polyester composite at different moments of the process. Analysis of this figure indicates that water migrates from the surface to center of the solid until saturation is reached. Absorption rate is higher at the first hour of the process, decreasing for longer times.

In order to predict which areas are more sensitive to thermo-hygro-mechanical stresses, it is important to know the moisture content inside the material as moisture may cause cracks, deformations and delaminations and, consequently, decrease the composites' mechanical properties.

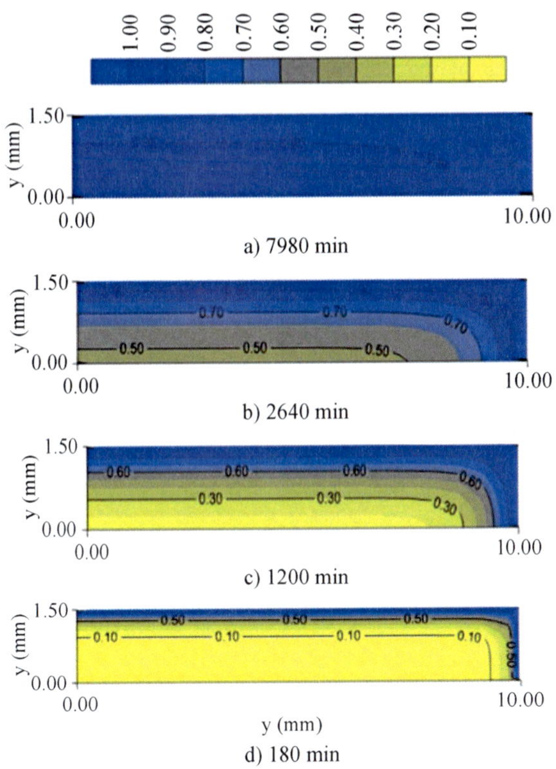

Fig. 4.18 Dimensionless moisture content distribution inside the macambira composite at different elapsed times (3 mm thickness, x = 5 mm, T = 25 °C)

4.2.5.3 CFD Simulation Results

(a) Water Sorption Process

Figure 4.19 illustrates a comparison between the predicted (Ansys CFX® software) and experimental average moisture content of caroá fiber-reinforced unsaturated polyester composite as a function of water exposure time (water absorption kinetics). Details of the experimental procedure are found in Chap. 3. Analyses of these curves show that a good agreement was obtained.

Transient behavior of the average moisture content of the polymer composite at different temperatures is illustrated in Fig. 4.20. Results indicate that, as expected, the water absorption of samples immersed in water at 70 °C is faster than those immersed at 25 and 50 °C. This behavior is attributed to the increased movement of water molecules within the solid due to the high temperature. It is observed that in the initial stages (first 50 h), the water absorption increases quickly and tends to continue increasing slowly until hygroscopic equilibrium is achieved. This effect is more pronounced for higher temperatures (T = 70 °C).

(b) Absorbed Water Distribution

Figures 4.21 and 4.22 show the moisture content spatial distribution inside the unsaturated polyester composites reinforced with caroá fiber at different planes: (x = 0, y, z), (x, y = 0, z), (x, y, z = 0) and elapsed times 4, 20, 44, 133 and 208 h of wetting. The reported results are those for water bath temperatures of 25 and 70 °C, respectively.

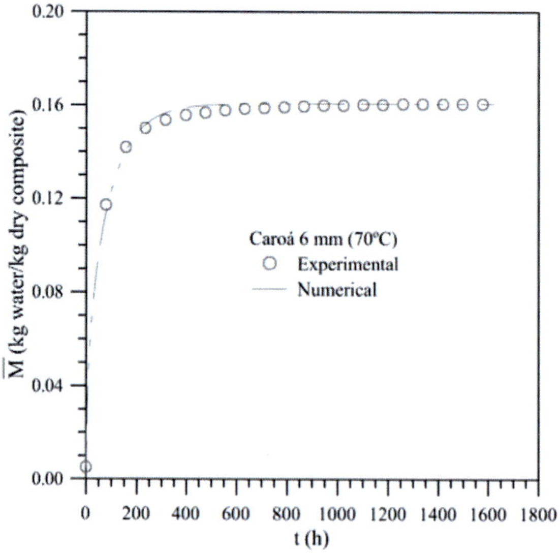

Fig. 4.19 Experimental and theoretical (proposed model) results of the average moisture content during the wetting process of unsaturated polyester composite reinforced with caroá fiber (thickness 6 mm and temperature 70 °C)

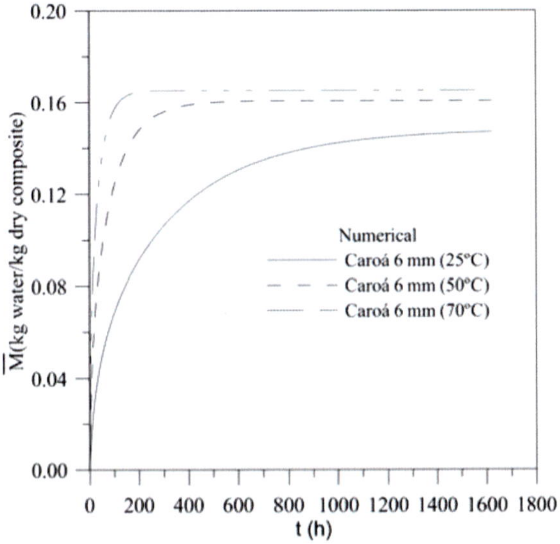

Fig. 4.20 Predicted water sorption kinetics of the polymer composites reinforced with caroá fibers at different water bath temperatures (6 mm thickness)

Analysis of these figures indicate that there is high moisture gradient in the composite near the vertex regions and that distribution of moisture content is not uniform with immersion time. In the areas close to the corners, water absorption is faster because these areas are in direct contact with water for the different sides of the samples. The water enters in the composite and generates higher moisture content gradients within the solid, decreasing as immersion time increases. Thus, at any point within the solid, the moisture content increases with time until the steady state condition (hygroscopic equilibrium condition) is reached.

It can be observed that, 20 h after the wetting process has started, moisture gradients at higher temperature are significantly higher than at lower temperature. This behavior is provoked by an increase in mass diffusion coefficient that is strogly dependent on water bath temperature. Then, the increase in moisture diffusion inside the composite with immersion time and test temperature can be attributed to a reduction in water density and viscosity as well as on the thermal expansion suffered by the matrix, which favors the penetration of water into the composite [32, 33].

4.2.5.4 Mass Diffusion Coefficient Estimation

The mechanism of diffusion and structure/diffusivity relationships, have not been extensively studied. Some authors have suggested that water diffusivity in polymer composites is due to morphology [34], local molecular mobility [35], free volume [36] or fiber distribution in the polymer matrix [37].

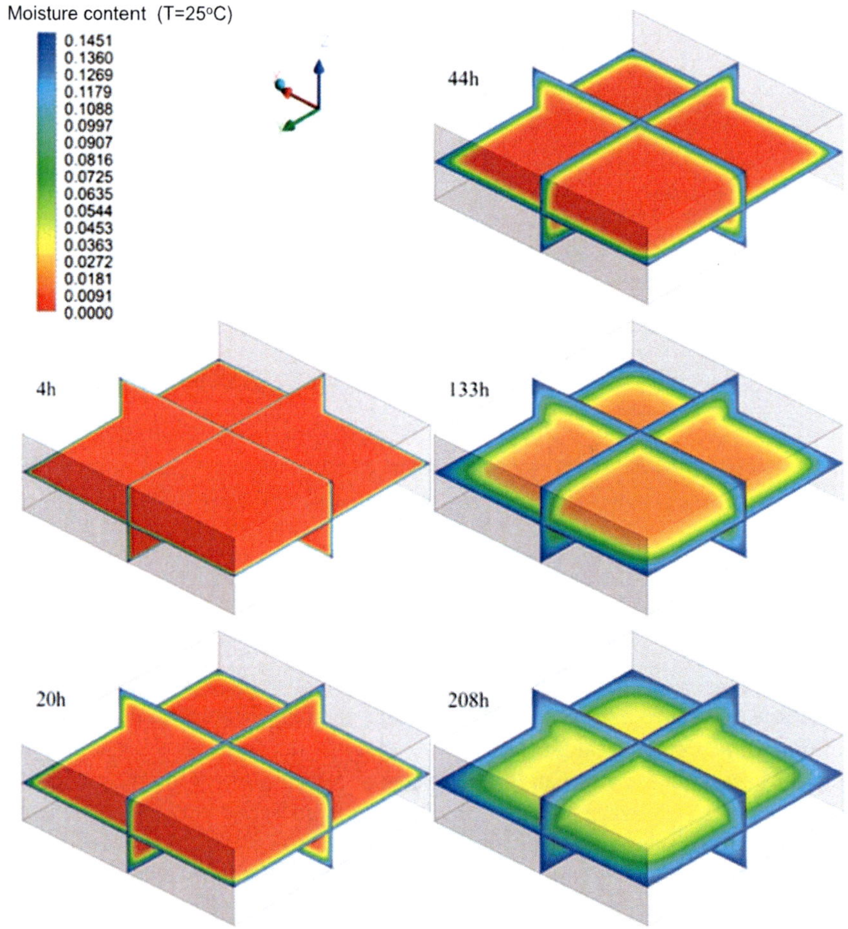

Fig. 4.21 Moisture content distribution inside the polymer composite reinforced with caroá fiber at different planes (x = 0, y, z), (x, y = 0, z) and (x, y, z = 0) and elapsed times (water bath temperature 25 °C)

According to Merdas et al. [38] apparent water diffusivity decreases with increasing hydrophilicity and it increases when temperature increases [31], showing that moisture diffusion rate within the solid is strongly affected by fiber/polymer/water interactions. Therefore, all factors mentioned above, including fiber content, thermal stability and shape of the composite, need to be investigated in order to fully understand moisture absorption mechanisms.

As previouly discussed, mass diffusion coefficients were estimated by fitting the predicted values to the experimental data obtained on average moisture content during the process of water absorption in unsaturated polyester/vegetable fiber composites.

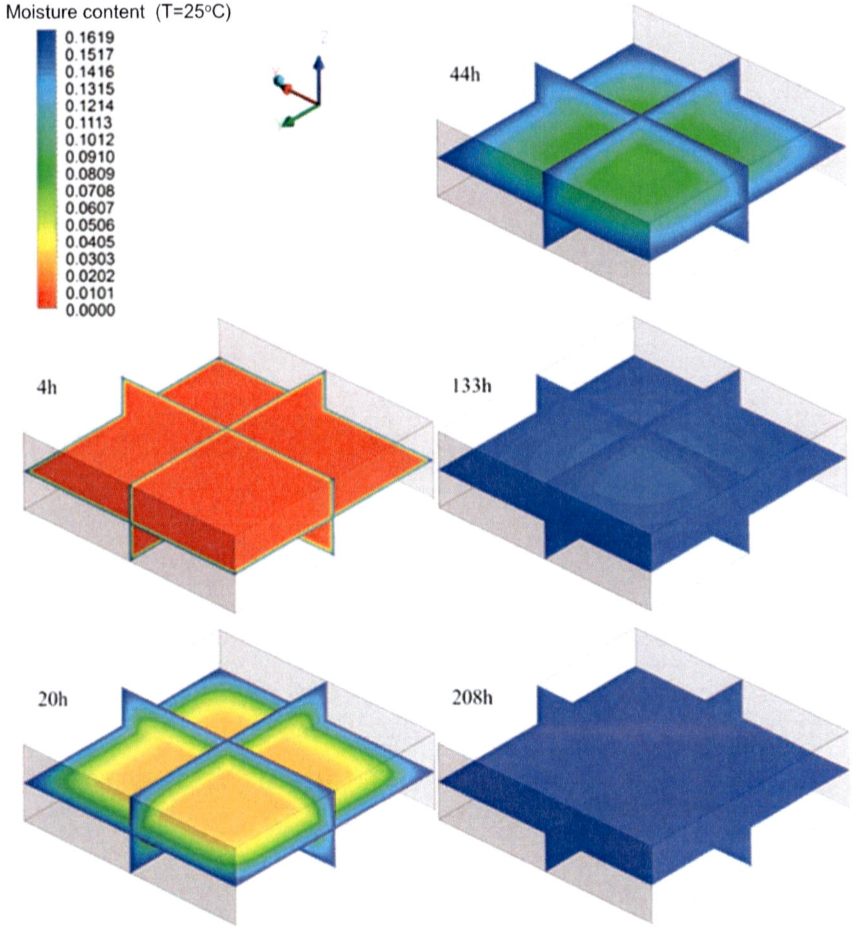

Fig. 4.22 Moisture content distribution inside the polymer composite reinforced with caroá fiber at different planes (x = 0, y, z), (x, y = 0, z) and (x, y, z = 0) and elapsed times (water bath temperature 70 °C)

Tables 4.1 and 4.2 present the value of the mass diffusion coefficient obtained for all experimental conditions presented in Chap. 3.

Analysis of these tables indicate that mass diffusion coefficient changes with moisture content and temperature for all cases investigated. This process parameter tends to decrease with increasing moisture content and to increase with increasing temperature for all samples. The use of moisture content dependent mass diffusivity led to numerical values that adjusted well to the experimental data, with smaller errors than those obtained when constant mass diffusivity was assumed.

The data obtained also showed that the mass diffusion coefficient is higher for thinner samples, which is attributed to the fact that the area to be coverd by the water

Table 4.1 Parameters of the Eq. 4.11 and variance estimated for all wetting experiments with caroá and mamcabira reinforced composites and pure polyester

Sample	$D_2 \times 10^{+12}$ m^2/s	C_2	ERMQ	$S^2 \times 10^{+3}$	Time (h)
Pure polyester 3 mm	2.8	–26.0	0.267	9.54	600
Caroá 3 mm 25 °C	2.2	–0.4	0.045	1.12	1095
Caroá 6 mm 25 °C	2.3	–0.1	0.093	1.94	1623
Caroá 3 mm 50 °C	8.0	–0.4	0.086	2.39	1022
Caroá 6 mm50 °C	8.5	–0.1	0.022	0.61	1022
Caroá 3 mm 70 °C	15.0	–0.4	0.035	1.35	796
Caroá 6 mm 70 °C	21.0	–0.1	0.043	1.65	796
Macambira 3 mm 25 °C	5.7	–1.32	0.1285	4.43	642
Macambira 6 mm 25 °C	2.9	2.00	0.1915	7.66	594
Macambira 3 mm 50 °C	11.2	–0.50	0.1254	4.18	687
Macambira 6 mm 50 °C	9.3	0.74	0.0513	1.71	687
Macambira 3 mm 70 °C	25.1	–0.30	0.1291	6.79	384
Macambira 6 mm 70 °C	21.4	0.58	0.0262	1.24	648
Sisal 3 mm 25 °C	3.04	0.00	0.0504	1.44	817
Sisal 6 mm 25 °C	1.68	0.00	0.1692	3.38	1328
Sisal 3 mm 50 °C	9.34	0.00	0.1073	3.07	817
Sisal 6 mm 50 °C	2.77	0.00	0.1698	4.72	848
Sisal 3 mm 70 °C	10.20	0.00	0.0493	1.41	817
Sisal 6 mm 70 °C	7.54	0.00	0.0529	1.47	848

during the humidifiation of thicker samples is bigger and the area/volume ratio is smaller. The data also showed that, for lower water bath temperatures, water sorption is dominated by water affinity to cellulose and to composite area while for higher water bath temperature, water absorption kinetics is dominated only by thermal effects. This implies that mass diffusion within the material is strongly dependent on temperature.

Based on the lower least square error and variance presented in Table 4.1, we conclude that the proposed model for the diffusion coefficient as a function of the average moisture content (Eq. 4.11) led to moisture content values that are in agreement with the experimental data.

It is well known that diffusion processes in systems with two or more components, such as polymer composites reinforced with vegetable fiber as they absorb water, depend on several factors such as volume fractions, nature, distribution and orientation of the fillers as well as on fiber/matrix interfacial characteristics [13, 39]. In general, the comparison with mass diffusion coefficients reported in the technical literature is very difficult for a number of different reasons that include models and calculation methods adopted, porosity, dimensions, temperature, compositions, and physical and chemical structures of the material [15, 40]. Table 4.3 summarizes some

Table 4.2 Range of the mass diffusion coefficients estimated by Eq. 4.11 and initial and final moisture content for all studied cases

Sample	Fiber total content (%)/T (°C)	$D_{initial} \times 10^{+12}$–$D_{final} \times 10^{+12}$ (m^2/s)	Initial and final moisture content (dry basis)
Pure polyester	0/28	2.80–2.31	0–0.01252
Caroá 3 mm	30/25	2.20–2.07	0–0.14488
Caroá 6 mm		2.30–2.26	0–0.14810
Caroá 3 mm	30/50	8.00–7.53	0–0.15162
Caroá 6 mm		8.50–8.36	0–0.16067
Caroá 3 mm	30/70	15.00–14.10	0–0.15609
Caroá 6 mm		21.00–20.65	0–0.16523
Macambira 3 mm	30/25	5.70–4.56	0–0.16810
Macambira 6 mm		2.90–3.84	0–0.14040
Macambira 3 mm	30/50	11.20–10.29	0–0.17020
Macambira 6 mm		9.30–10.46	0–0.15890
Macambira 3 mm	30/70	25.10–23.77	0–0.18090
Macambira 6 mm		21.40–23.51	0–0.16180
Sisal 3 mm	44.6/25	3.04–3.04	0–0.1468
Sisal 6 mm		1.68–1.68	0–0.1227
Sisal 3 mm	44.6/50	9.34–9.34	0–0.1496
Sisal 6 mm		2.77–2.77	0–0.01242
Sisal 3 mm	44.6/70	10.20–10.20	0–0.1504
Sisal 6 mm		7.54–7.54	0–0.1246

mass diffusity values reported in the literature for different materials, sample shape, and experimental conditions.

A reasonable agreement is observed when the mass diffusivity values of different composites obtained in this study are compared with those reported in similar systems. Since the diffusive model was used in all works reported in Table 4.3, the difference between the values may be attributed mainly to composite variety; geometry assumptions; different equilibrium moisture content; composite physical structure and expansion effects caused by water absorption.

As final comment, despite of the extended use of the Fick's second law to study diffusion process, due to several reasons such as experience of the researchers, available equipments, and available exact solutions of the diffusion equation for simple geometries, it fails to address the complex structure of the polymer composite absorption process. The reason for this failure is the different water absorption capacity of the polymer matrix and the natural fiber, and their iteractions inside the composite, especialy in the final stages of the process. Thus, new or improved mathematical modelis to adequately predict water migration in poymer composites when exposed to moist environment for long times have been proposed in the literature. This topic is addressed in detail in the next chapter of this book.

Table 4.3 Moisture diffusivity in polymer composites as reported in the literature

Composite sample (Shape)	Time (h)	Dimension (mm)	T (°C)	$D \times 10^{+13}$ (m^2/s)	References
Carbon-glass fiber/Epoxy (Rod)	5300	66.5 × 9.53	40	1.42	[31]
Carbon-glass fiber/Epoxy (Rod)	5300	66.5 × 9.53	60	3.84	[31]
Carbon-glass fiber/Epoxy (Rod)	5300	66.5 × 9.53	90	27.00	[31]
Sisal fiber/Polypropylene (Rectangular prism)	166	25 × 10 × 2.50	28	0.36	[29]
Sisal fiber/Polypropylene (Rectangular prism)	166	25 × 10 × 2.50	50	0.56	[29]
Sisal fiber/Polypropylene (Rectangular prism)	166	25 × 10 × 2.50	28	0.69	[29]
Jute-glass fiber/Polyester (Rectangular prism)	1249	20 × 20 × 2.82	28	16.50	[41]
Jute fiber/Polyester (Rectangular prism)	500	20 × 20 × 2.30	28	18.00	[42]
Kenaf fiber/Polyester (Rectangular prism)	911	127 × 12.5 × 3.2	25	25	[43]
Kenaf fiber/Polyester (Rectangular prism)	911	127 × 12.5 × 3.2	50	108	[43]
Sisal-Cotton fiber/Polyester (Infinite plate)	450	20 × 20 × 3.0	30	0.17	[44]
Sisal-Cotton fiber/Polyester (Infinite plate)	450	20 × 20 × 3.0	60	0.67	[44]
Ramie-Cotton fiber/Polyester (Infinite plate)	450	20 × 20 × 3.0	30	0.33	[44]
Ramie-Cotton fiber/Polyester (Infinite plate)	450	20 × 20 × 3.0	60	1.51	[44]
Jute-Cotton fiber/Polyester (Infinite plate)	450	20 × 20 × 3.0	30	0.05	[44]
Jute-Cotton fiber/Polyester (Infinite plate)	450	20 × 20 × 3.0	30	0.55	[44]

References

1. Farias, V.S.O., Silva, W.P., Silva, C.M.D.P.S., Delgado, J.M.P.Q., Farias Neto, S.R., Lima, A.G.B.: Transient diffusion in arbitrary shape porous bodies: Numerical analysis using boundary-fitted coordinates. In: Delgado, J.M.P.Q., de Lima, A.G.B, Silva, M.V. (eds.) Numerical Analysis of Heat and Mass Transfer in Porous Media, vol. 27, pp. 85–119. Springer, Heidelberg (2012)
2. Ellis, B.E., Found, M.S.: The effects of water absorption on a polyester/chopped strand mat laminate. Compos. **14**(3), 237–243 (1983)
3. Han, K.S., Koutsky, J.: Effect of water on the interlaminar fracture behaviour of glass fibre-reinforced polyester composites. Compos. **14**, 67–70 (1983)
4. Camino, G., Luda, M.P., Polishchuk, A.Y., Revellino, M., Blancon, R., Merle, G., Martinez-Veja, J.J.: Kinetic aspects of water sorption in polyester resin/glass-fibre composites. Compos. Sci. Technol. **57**, 1469–1482 (1997)
5. Choi, H.S., Ahn, K.J., Nan, J.-D., Chun, H.J.: Hygroscopic aspects of epoxi/carbon fiber composite laminates in aircraft environments. Compos. Part A: Appl. Sci. Manuf. **32**, 709–720 (2001)
6. Srihari, S., Revathi, A., Rao, R.M.V.G.K.: Hygrothermal effects on RT-cured glass-epoxi composites in immersion environments. Part A: Moisture absorption characteritics. J. Reinf. Plast. Compos. **21**, 11, 983–991 (2002)
7. Yao, J., Ziegmann, G.: Water absorption behavior and its influence on properties of GRP pipe. J. Compos. Mater. **41**(8), 993–1008 (2007)
8. Najafi, K.S., Kiaefar, A., Hamidinia, E., Tajvidi, M.: Water absorption behavior of composites from sawdust and recycled plastics. J. Reinf. Plast. Compos. **26**(3), 341–348 (2007)
9. Katzman, H.A., Castaneda, R.M., Lee, H.S.: Moisture diffusion in composite sandwich structures. Compos. Part A: App. Sci. Manuf. **39**(5), 887–892 (2008)
10. Czél, G., Czigány, T.: A study of water absorption and mechanical properties of glass fiber/polyester composite pipes — Effects of specimen geometry and preparation. J. Compos. Mater. **42**(26), 2815–2827 (2008)
11. Chateauminois, A., Vicent, L., Chabert, B.E., Soulier, J.P.: Study of the interfacial degradation of a glass-epoxi composite during hygrothermal ageig using water diffusion measurements and dynamic mechanical thermal analysis Polym. **35**(22), 4766–4774 (1994)
12. Pavan, R.M.V., Saravanan, V., Dinesh, A.R., Rao, Y.J., Srihari, S., Revathi, A.: Hygrothermal effects on painted and unpaited glass/epoxy composites- part a: moisture absorption characteristics. J. Reinf. Plast. Compos. **20**(12), 1036–1046 (2001)
13. Bao, L.R., Yee, A.F.: Moisture diffusion and hygrothermal aging in bismaleimide matrix carbon fiber composites-part I: uni-weave composites. Compos. Sci. Technol. **62**, 2099–2110 (2002)
14. Cavalcanti, W.S., Carvalho, L.H., Lima, A.G.B.: Water diffusion in unsaturated polyester composite reinforced by a hybrid jute/glass fabric: modeling and simulation. Rev. Matér. **10**(1), 14–23 (2005) (In Portuguese)
15. Cavalcanti, W.S.: Polyester/hybrid jute-glass fabric composites: mechanical characterization and water sorption simulation Doctorate Thesis. Federal University of Campina Grande, Paraíba, Brazil, Process Engineering (2006).(In Portuguese)
16. Santos, D.G.: Thermo-hydric study and mechanical characterization of polymeric matrix composites reinforced with vegetable fiber: 3D simulation and experimentation. Doctoral Thesis in Process Engineering, Federal University of Campina Grande, Campina Grande, Brazil (2017)
17. Santos, D.G., Lima, A.G.B., Costa, P.S., Lima, E.S., Moreira, G., Nascimento, J.J.S.: Water absorption in sisal fiber reinforced-polymeric matrix composites: three-dimensional simulations and experiments. Diffus. Found. **20**, 143–154 (2018)
18. Melo, J.B.C.A.: Water absorption in fiber reinforced polymer composites of pineapple leaf: modeling and simulation. Doctorate Thesis in Process Engineering. Federal University of Campina Grande, Campina Grande, Brazil (2014) (In Portuguese)

19. Carvalho, L.H., Canedo, E.L., Neto, S.F., Lima, A.G.B.: Moisture transport process in vegetable fiber composites: theory and analysis for technological applications. In: Delgado, J.M.P.Q. (ed.) Industrial and Technological Applications of Transport in Porous Materials, pp. 37–62. Springer, Berlin (2013)
20. Luikov, A.V.: Analytical Heat Diffusion Theory. Academic, London (1968)
21. Gebhart, B.: Heat Conduction and Mass Diffusion. McGraw-Hill Inc., New York (1993)
22. Patankar, S.V.: Numerical Heat Transfer and Fluid Flow. Hemisphere Publishing Corporation, New York (1980)
23. Maliska, C.R.: Computational Heat Transfer and Fluid Mechanics. LTC, Rio de Janeiro (2004).(In Portuguese)
24. Versteeg, H.K.; Malalasekera, W.: An Introduction to Computational Fluid Dynamics: the finite volume method. Pearson Education Limited, Harlow, England (1995)
25. Figliola, R.S., Beasley, D.E.: Theory and Design for Mechanical Measurements. Wiley, New York (1995)
26. Silva, C.J.: Water Absorption in Composite Materials of Vegetal Fiber: Modeling and Simulation via CFX. Master Dissertation in Mechanical Engineering, Federal University of Campina Grande, Campina Grande, Brazil (2014)
27. Silva, C.J., Barbosa de Lima, A.G., Silva, E.G, Andrade, T.H.F., MELO, R.Q.C.: Water Absorption in Caroá-Fiber Reinforced Polymer Composite at Different Temperatures: A Theoretical Investigation. Diffusion Foundations, vol. 10, pp. 16–27 (2017)
28. Silva, C.J., Andrade, T.H.F., Silva, E.G., Lima, A.G.B.: Water absorption in composites reinforced with Caroa- fiber fabrics: modeling and simulation via ANSYS CFX®. Defect Diffus. Forum. **353**, 84–89 (2014)
29. Joseph, P.V., Rabello, M.S., Mattoso, L.H.C., Joseph, K., Thomas, S.: Environmental effects on the degradation behaviour of sisal fibre reinforced polypropylene composites. Compos. Sci. Technol. **62**, 1357–1372 (2002)
30. Chow, C.P.L., Xing, X.S., Li, R.K.Y.: Moisture absorption studies of sisal fibre reinforced polypropylene composites compos. Sci. Technol. **67**, 306–313 (2007)
31. Tsai, Y.I., Bosze, E.J., Barjasteh, E., Nutt, S.R.: Influence of hygrothermal environment on thermal and mechanical properties of carbon fiber/fiberglass hybrid composites. Compos. Sci. Technol. **69**, 432–437 (2009)
32. Nóbrega, M.M.S.: Polyester matrix composite with caroá fibers (Neoglaziovia variegata): Mechanical charcterization and water sorption, Doctorate Thesis, Process Engineering, Federal University of Campina Grande, Paraíba, Brazil (2007), (In Portuguese)
33. Nóbrega, M.M.S., Cavalcanti, W.S., Carvalho, L.H., Lima, A.G.B.: Water absorption in unsaturated polyester composites reinforced with caroá fiber fabrics: modeling and simulation. Mat.-wiss. u.Werkstofftech. **41**, 300–305 (2010)
34. Gupta, V.B., Drzal, J.L., Rich, M.J.: The physical basis of moisture transport in a cured epoxy resin system. J. Appl. Polym. Sci. **30**(11), 4467–4493 (1985)
35. Lagouvardos, P.E., Pissis, P., Kyritsis, A., Daoukaki, D.: Water sorption and water-induced molecular mobility in dental composite resins. J. Mater. Sci. Mater. Med. **14**(9), 753–759 (2003)
36. Duda, J.L., Vrentas, J.S., Ju, S.T., Liu, H.T.: Prediction of diffusion coefficients for polymer-solvent systems. AIChE J. **28**, 279–285 (1982)
37. Wang, W., Sain, M., Cooper, P.A.: Study of moisture absorption in natural fiber plastic composites. Compos. Sci. Technol. **66**, 379–386 (2006)
38. Merdas, I., Thominette, F., Tcharkhtchi, A., Verdu, J.: Factors governing water absorption by composite matrices. Compos. Sci. Technol. **62**, 487–492 (2002)
39. Marcovich, N.E., Reboredo, M.M., Aranguren, M.I.: Moisture diffusion in polyester - woodflour composites. Polym. **40**(26), 7313–7320 (1999)
40. Lima, A. G. B.: Diffusion phenomena in prolate spheroidal solids: Studied case: Drying of banana. Doctorate Thesis, Mechanical Engineering, State University of Campinas, Campinas, Brazil (1999) (In Portuguese)

41. Silva, W. P., Silva C. M. D. P. S., Farias, V. S. O., Lima A. G. B.: Effect of the geometry on the description of the water absorption by composite materials using diffusion models. Mat.-wiss. u.Werkstofftech. 42, 8, 747–752 (2011)
42. Cavalcanti, W. S., de Lima, A. G. B., Carvalho, L. H.: Water sorption in unsaturated polyester composites reinforced with jute and jute/glass fiber fabrics: Modeling, simulation and experimentation. Polímeros: Ciência e Tecnologia 20, 1, 78–83 (2010) (In Portuguese)
43. Osman, E., Vakhguelt, E., Sbarski, I., Mutasher, S.: Water absorption behavior and its effect on the mechanical properties of kenaf natural fiber unsaturated polyester composites. In: Proceedings of the 18th International Conference on Composite Materials, Edinburgh, Scotland (2009)
44. Alsina, O.L.S.; Carvalho, L.H.; Ramos Filho, F.G.; d'Almeida, J.R.M.: Immersion temperature effects on the water absorption behavior of hybrid lignocellulosic fiber reinforced-polyester matrix composites. Polym. – Plast. Technol. Eng. **46**(5), 515–520 (2007)

Chapter 5
Langmuir-Type Model Analysis

In this chapter, based on the rigorous theory of anomalous diffusion through fibrous porous media (Langmuir-type model), its advantagens, limitations, and different approaches (analytical and numerical), we present the correct prediction for the water absorption process in vegetable fiber-reinforced polymer composites. In the macroscopic and advanced mathematical modeling, both fiber and polymer are considered a homogeneous mixture having water molecules in the free and entrapped states inside the material, and the effect of swelling is neglected. Herein, different 1D and 3D simulation results of the average moisture content, moisture content as well as free and entrapped water molecules concentration distribution at different times are presented and analyzed. Applications have been focused to caroá fiber- reinforced polyester composites and other arbitrary cases.

5.1 Fundamentals

The effects of moisture absorption on the physical and chemical properties of polymer composite materials have received wide attention, not only due to the durability of these materials in operation but also because of their wide field of application.

As already mentioned, the properties of composite materials depend on the behavior of the matrix, the reinforcement and the fiber/matrix interface. Absorption of water molecules in materials leads to degradation of the interface and swelling of the matrix, generating internal stresses and decreasing mechanical properties. Decreasing properties are intrinsically associated to the moisture diffusion process in composite materials and therefore, in order to predict long-term behavior, in-service failure and useful life estimation [1], it is essential to know the water absorption rate.

Diffusion models must reflect the moisture sorption process in different ways, considering certain additional phenomena that can occur inside the material. Thus,

J. M. P. Q. Delgado et al., *Moisture Transport in Polymer Composite Materials*,
Advanced Structured Materials 160, https://doi.org/10.1007/978-3-030-77826-2_5

each specific case must be analyzed not only based on the results of a good approximation between the experimental data and those predicted by the equations, but also by the physical interpretation of the water sorption process included in the models.

In some cases, the water absorption kinetics is described assuming that water absorption follows the classical treatment given by Fick's second law of diffusion, that is, that the diffusion process is driven by the water concentration gradient between the medium and the material and continues until hygroscopic equilibrium is reached. In others, this model is not well suited [2, 3], because the material presents an anomalous water diffusion behavior, which implies in the existence of different stages of sorption until the final mass balance is reached (hygroscopic equilibrium condition). In this case, single-phase models cannot easily describe this kind of diffusion, and one needs other more sophisticated models to describe the behavior of the material under these physical conditions [4]. The Langmuir-type model is appropriate to predict the non-Fickian diffusion process in this situation [5].

The Langmuir-type model explains moisture sorption by assuming that water exists simultaneously in two phases: free (unbound) and entrapped (bound), during the process. In this model, water molecules in the free phase are adsorbed (bound) with a probability α per unit time, and the entrapped water molecules can leave the connected state with a probability β per unit time. Thus, the diffusion process is described by the classical diffusion equation, which is modified to take into account the two phases of the moisture inside the material.

Figure 5.1 illustrates the moisture content transient behavior predicted by Fick´s model and Langmuir-type model. According to the Langmuir-type model, during the water absorption process, moisture migration is controlled by two phenomena: dispersion of free water molecules due to random molecular motion and trapping of water molecules due to interaction between the free water molecules and porous within the polymer composite.

In Fig. 5.1, the pseudo-equilibrium point represents the point at which the moisture migration rate drops dramatically.

The Langmuir-type model is well described in Glaskova et al. [6], Carter and Kilber [5], Bonniau and Bunsell [1], Grace and Altan [2], Cotinaud et al. [7] and Apicella et al. [8].

According to Carter and Kilber [5], the diffusive characteristics of the Langmuir-type model are correlated to the simplest form of neutron transport theory, while that the characteristics associated to free and entrapped water molecules are related to the classical Langmuir´s adsorption isotherms theory.

According to these authors, composite anisotropy (structural heterogeneity), swelling, size and distribution of micropores (microvoids), mechanical loading, rearrangement of the polymer network (long-term relaxation), or some other phenomenon not yet fully understood are responsible for the intensity of the physical or chemical interactions at the microscopic level between polar water molecules and molecular groups of the polymer chains (molecular binding). A detailed discussion on the applications of this model are presented in the next section.

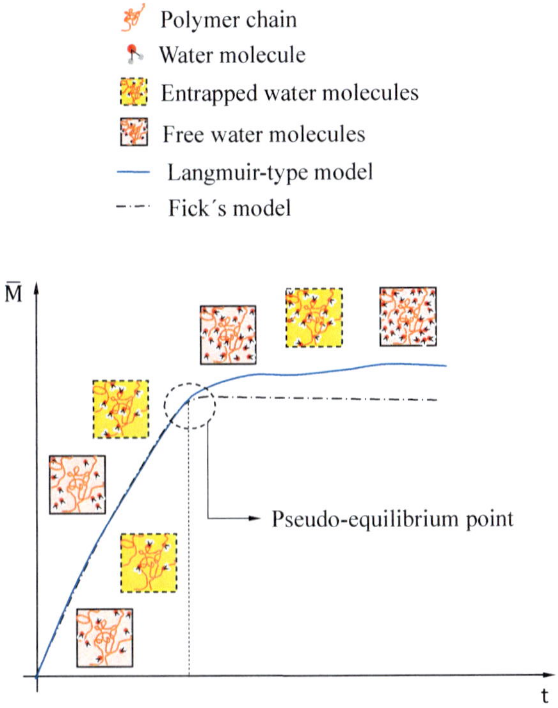

Fig. 5.1 Typical moisture absorption kinetics predicted by Langmuir-type model and Fick´s model

5.2 Moisture Absorption by Langmuir-Type Model

5.2.1 The Physical Problem

For the analysis of the physical problem studied here, consider a dry polymer composite of parallelepiped geometry, at low temperature, suddenly immersed in a stationary, heated, saturated fluid (water). As the surrounding fluid heats the dry porous media, heat penetrates into the solid (as the result of a temperature difference) and moisture migrates into the solid by diffusion from the surface. Figure 5.2 illustrates the problem treated here. In this figure Lx, Ly, and Lz are the distances from the composite surface to the maximum level of water in the container.

5.2.2 The General Mass Diffusion Equation

In the Langmuir-type model, the non-Fickian moisture absorption behavior can be explained quantitatively by assuming that moisture absorption occurs in the presence of two simultaneous stages, one being the free water stage and the another the entrapped water stage [5]. The following mass transfer equations describe this model:

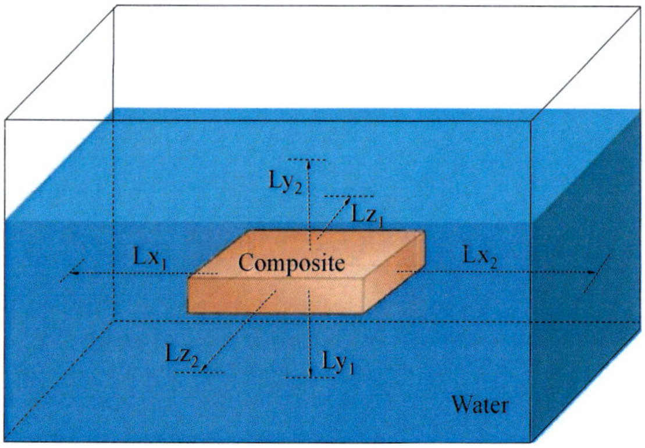

Fig. 5.2 Squematic of the composite immersed in a stationary fluid and the distance of the material to borders of the fluid domain

$$\frac{\partial C}{\partial t} = \nabla \cdot (D\nabla C) - \frac{\partial S}{\partial t} \tag{5.1}$$

$$\frac{\partial S}{\partial t} = \lambda C - \mu S \tag{5.2}$$

where: C is the concentration of the free solute diffusing into the material; S is the concentration of the entrapped solute; D is the mass diffusion coefficient (free solute molecules); t is the time; λ is the probability of a free solute molecule is entrapped inside the solid and μ is the probability that an entrapped solute molecule becomes free.

Careful analysis of Eq. (5.1), indicates that if $\lambda = 0$ or $\mu >> \lambda$, it is reduced to the simple diffusion theory (Fick´s second law of diffusion).

5.2.3 The Mass Diffusion Equation: 3D Approach in Cartesian Coordinates

For a three-dimensional and transient approach, based on the considerations adopted, the Langmuir model, in Cartesian Coordinates, can be written as:

$$\frac{\partial C}{\partial t} = \frac{\partial}{\partial x}\left(D\frac{\partial C}{\partial x}\right) + \frac{\partial}{\partial y}\left(D\frac{\partial C}{\partial y}\right) + \frac{\partial}{\partial z}\left(D\frac{\partial C}{\partial z}\right) - \frac{\partial S}{\partial t} \tag{5.3}$$

$$\frac{\partial S}{\partial t} = \lambda C - \mu S \tag{5.4}$$

In order to solve Eqs. 5.3 and 5.4, the following initial and boundary conditions were used:

(a) ***Initial conditions***: if the solid is completely dry at the beginning of the process, then one can write:

$$C = S = 0; \begin{cases} -R_1 < x < R_1 \\ -R_2 < y < R_2 \\ -R_3 < z < R_3 \end{cases} \tag{5.5}$$

(b) ***Boundary conditions:*** the variation in the concentration of the solute in the fluid medium on the surface of the solid is equal to the diffusive flux of solute into the material. Then, one can write:

$$Lx_1 \frac{\partial C}{\partial t} = -D\frac{\partial C}{\partial x}; \begin{cases} x = -R_1 \\ t > 0 \end{cases} \tag{5.6}$$

$$Lx_2 \frac{\partial C}{\partial t} = -D\frac{\partial C}{\partial x}; \begin{cases} x = +R_1 \\ t > 0 \end{cases} \tag{5.7}$$

$$Ly_1 \frac{\partial C}{\partial t} = -D\frac{\partial C}{\partial y}; \begin{cases} x = -R_2 \\ t > 0 \end{cases} \tag{5.8}$$

$$Ly_2 \frac{\partial C}{\partial t} = -D\frac{\partial C}{\partial y}; \begin{cases} x = +R_2 \\ t > 0 \end{cases} \tag{5.9}$$

$$Lz_1 \frac{\partial C}{\partial t} = -D\frac{\partial C}{\partial z}; \begin{cases} x = -R_3 \\ t > 0 \end{cases} \tag{5.10}$$

$$Lz_2 \frac{\partial C}{\partial t} = -D\frac{\partial C}{\partial z}; \begin{cases} x = +R_3 \\ t > 0 \end{cases} \tag{5.11}$$

Once C and S are determined at any point inside the material, it is possible to calculate the total moisture content present in the material at any position and instant of time. Therefore, the total moisture content is given by the sum of the concentrations of C and S, as follows:

$$M(x,y,z,t) = C(x,y,z,t) + S(x,y,z,t) \tag{5.12}$$

It follows from Eq. 5.12 that the average moisture content of the solid at any time of the process is given by:

$$\overline{M} = \frac{1}{V}\int_V M(x,y,z,t)dV \tag{5.13}$$

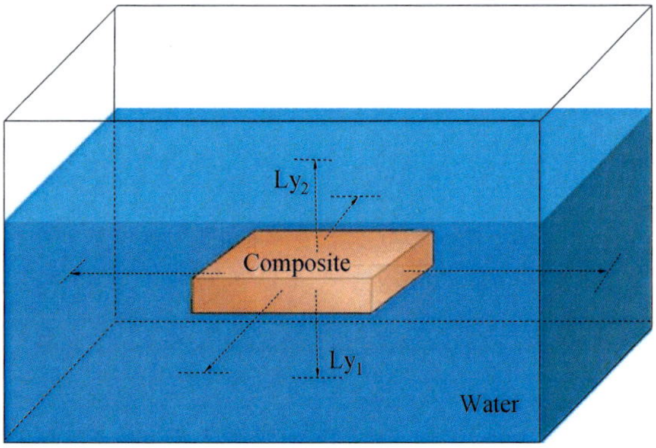

Fig. 5.3 Geometrical representation of the one-dimensional physical problem

in which V is the total volume of the solid and dV = dxdydz is the volume of an infinitesimal sample of the porous solid.

5.2.4 *The Water Absorption Process: 1D Approach in Cartesian Coordinates*

5.2.4.1 The Physical Problem

For the physical model, a porous plate with thickness $2R_2 = 2a$, immersed in a fluid solution (water) of height $(Ly_1 + 2a + Ly_2)$, as illustrated in Fig. 5.3 was considered.

Considering that $R_1 = R_3 >> R_2$, water absorption can be analysed just in the y direction and the following assumptions were made: the material is homogeneous and isotropic; the mass diffusion coefficient is constant; the solid is axisymmetric; the process is transient; the dimensions of the material during the diffusion process do not change; the capillary transport through the solid is negligible; mass generation inside the solid is neglected; the solid is totally dry at the beginning of the process and the solid is in equilibrium with the surrounding medium at the surface (equilibrium boundary condition).

5.2.4.2 The Mass Diffusion Equation

In the Langmuir-Type model, the anomalous moisture absorption can be quantitatively determined by assuming that absorbed moisture consists of a mobile phase and a bound phase. The model considers the interaction between the polar water molecules and the resin molecular groups, predicting the existence of free and bound molecules within the polymer network. This can be taken into account by adding a new parameter to the classical Fick's equation [9]. Considering the assumptions already cited, the Langmuir equation, written in Cartesian coordinates in a one-dimensional approach, is described as follows:

$$\frac{\partial C}{\partial t} = D\frac{\partial^2 C}{\partial y^2} - \frac{\partial S}{\partial t} \tag{5.14}$$

where,

$$\frac{\partial S}{\partial t} = \lambda C - \mu S \tag{5.15}$$

For the proposed problem the following initial and boundary conditions were considered:

- Initial condition:

$$S = C = 0, -a < y < a, t = 0 \tag{5.16}$$

- Boundary condition:

$$Ly_1\frac{\partial C}{\partial t} = -D\frac{\partial C}{\partial y} \quad y = -a, t > 0 \tag{5.17}$$

$$Ly_2\frac{\partial C}{\partial t} = -D\frac{\partial C}{\partial y} \quad y = +a, \ t > 0 \tag{5.18}$$

where Ly_1 and Ly_2 represent the distance between the solid surface and the bottom and top of the water tank, respectively. According to Eqs. (5.17) and (5.18), it is assumed that the rate of solute that leaves the solution is equal to the diffusive flux of solute at the surface of the plane sheet (see Fig. 5.3).

5.2.4.3 Solution Techniques: 1D Approach

Analytical Solution

Based on the works of Carter and Kibler [5] and Crank [10], Santos et al. [3] present the exact solution for the Eqs. (5.14) and (5.15) using the method of Laplace Transform. The application of the Laplace Transform consists in converting a partial differential equation in an ordinary differential equation which can be solved more easily. After this procedure, the inverse Laplace transform is calculated to get the original function of the problem that represent the solution of the governing equations [11, 12].

In order to obtain the exact solution of the physical problem, the model was simplified considering a porous plate with thickness $2R_2 = 2a$, immersed in a fluid (water) of height $(2L + 2a)$, where $Ly_2 = Ly_1 = L$ (Fig. 5.3).

The final equation for the concentration of free solute inside the solid during the water absorption process is obtained by considering the boundary conditions, the domain of the functions and with the use of necessary simplifications. This equation can be written as:

$$C(y,t) = \frac{LC_e}{L + (R+1)a} + \sum_{n=1}^{\infty} \frac{C_e \cos(k_n y) Exp(p_n t)}{\cos(k_n a)\left\{1 + \left[1 + \frac{\mu\lambda}{(p_n+\mu)^2}\right]\left[\frac{Lp_n^2 a}{2D^2k_n^2} + \frac{p_n}{2Dk_n^2} + \frac{a}{2L}\right]\right\}} \tag{5.19}$$

where p_n and k_n are eigenvalues and C_0 represent the initial solute concentration.

The final equation for the concentration of solute entrapped on the solid is written as:

$$S(y,t) = \left(\frac{\lambda}{\mu}\right)\frac{LC_e}{L + (R+1)a} + \sum_{n=1}^{\infty}\left(\frac{\lambda}{p_n+\mu}\right) \frac{C_e \cos(k_n y) Exp(p_n t)}{\cos(k_n a)\left\{1 + \left[1 + \frac{\mu\lambda}{(p_n+\mu)^2}\right]\left[\frac{Lp_n^2 a}{2D^2k_n^2} + \frac{p_n}{2Dk_n^2} + \frac{a}{2L}\right]\right\}} \tag{5.20}$$

The total moisture content inside the material in a specific position x and instant t is obtained from the sum of the amount of free solute and the amount of solute entrapped according with the following equation:

$$M = S + C \tag{5.21}$$

Based on Eq. (5.21), the average moisture content of the solid at different moments of the water uptake can be computed as follows:

$$\overline{M} = \frac{1}{V}\int_V M dV \tag{5.22}$$

or yet,

$$\overline{M} = \frac{1}{2a}\int_{-a}^{a} M(y,t)dy \tag{5.23}$$

where V is the volume of the solid.

From Eq. (5.23), the average moisture content of the solid at different water uptake times can be computed as follows:

$$\frac{\overline{M}}{M_e} = 1 - \sum_{n=1}^{\infty} \frac{(1+\alpha)e^{p_n t}}{1 + \left[1 + \frac{\mu\lambda}{(p_n+\mu)^2}\right]\left[\frac{Lp_n^2 a}{2D^2k_n^2} + \frac{p_n}{2k_n^2 D} + \frac{a}{2L}\right]} \tag{5.24}$$

where:

$$\alpha = \frac{L}{(R+1)a} \tag{5.25}$$

$$M_e = \frac{LC_e}{(1+\alpha)a} \tag{5.26}$$

In Eq. (5.24), $\overline{M}$ corresponds to the total amount of solute, both free and immobilized to diffusion at a given time t, M_e corresponds to the amount of moisture at equilibrium, obtained after an infinite time, and $R = {}^{\lambda}/_{\mu}$. The terms p_n and k_n together form pairs of eigenvalues and aim to refine the approximate calculation and the results. Thus, the higher the number of eigenvalues, the more accurate the analytical results obtained. They correspond to non-zero roots of the Eq. (5.27) that are originated from applying Eqs. (5.19) in (5.18).

$$\frac{Lp}{D} = k\tan(k \times a) \tag{5.27}$$

In Eq. (5.27), the values of k are given by:

$$k^2 = -\frac{p}{D}\left(\frac{p+\mu+\lambda}{p+\mu}\right) \tag{5.28}$$

For a 3D physical situation, the analytical solution can be obtained as the product of the analytical solution for three infinite plates. In this case, each of the plates must have thickness equal to $2R_1$, $2R_2$ and $2R_3$. Special care must be given in the determination of the eigenvalues p and k, which are different for each of the x, y and z diretions [2, 13–15].

Numerical Solution

The numerical solution of a partial differential equation basically consists of two steps: (a) discretizing the physical domain under study in several sub domains and (b) transforming the governing equation into a linear algebraic equation in the discretized form applied to each sub domain contained in the solid under investigation. After these procedures, the result is a set of linear algebraic equations whose solution provides the distribution of the potential unknown inside the domain and in time.

Herein, the finite-volume method for numerical solution of the governing equation [16–18] was used. For the discretization of Eqs. (5.14) and (5.15), the continuous solid of a thickness <u>2a</u> was subdivided in (np-2) control volumes, as shown in Fig. 5.4.

In Fig. 5.4 each control volume has thickness Δx; S, P and N represent nodal points, while <u>s</u> and <u>n</u> represent the left and right faces of the control volume P,

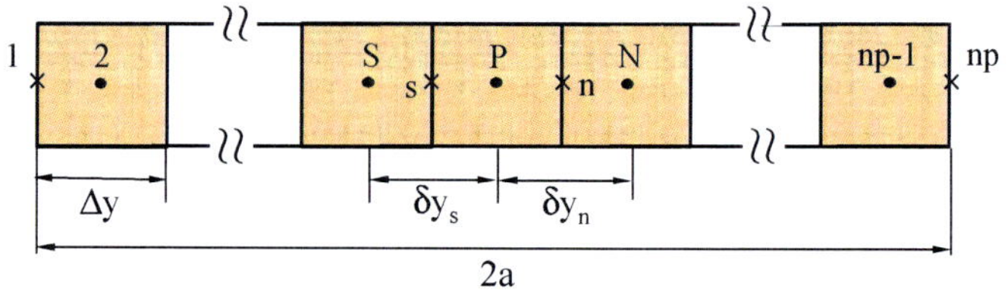

Fig. 5.4 Representation of the one-dimensional simulation domain with (np-2) control volumes

respectively; δy_s and δy_n represent the distance between nodal point P and nodal points S and N, respectively.

(a) **Solution for the concentration of free solute**

- ***Internal points***

In the finite volume method, discretization is done by integrating all terms of Eq. (5.14) in volume and time. Thus, after integration and using a fully implicit formulation, rearranging the terms in the linearized discrete algebraic form at P, leads to:

$$\mathrm{A_P C_P = A_N C_N + A_S C_S + A_P^o C_P^o + B_P^C} \tag{5.29}$$

where,

$$\mathrm{A_P} = \left(\frac{\Delta x}{\Delta t} + \frac{\mathrm{D_n}}{\delta x_{\mathrm{n}}} + \frac{\mathrm{D_s}}{\delta x_{\mathrm{s}}} + \lambda \Delta \mathrm{x}\right) \tag{5.30}$$

$$\mathrm{A_N} = \frac{\mathrm{D_n}}{\delta \mathrm{y_n}} \tag{5.31}$$

$$\mathrm{A_S} = \frac{\mathrm{D_s}}{\delta \mathrm{y_s}} \tag{5.32}$$

$$\mathrm{A_P^o} = \frac{\Delta \mathrm{y}}{\Delta t} \tag{5.33}$$

$$\mathrm{B_P^C} = \mu \Delta \mathrm{y} \mathrm{S_P^o} \tag{5.34}$$

where the coefficients A_P, A_N, and A_S are the conductance between the nodal points P and their corresponding neighbors. The term A_P^o represents the influence of the value of C on the value of C at t prior to that at the current time t. Equation (5.29) is valid for all internal points of the domain except for the boundary and symmetry points.

- ***Bottom boundary points***

In this case, the boundary flux is replaced by the existing boundary condition (Eq. 5.17). By doing this, the linear discretized equation for the bottom boundary points is:

$$A_P C_P = A_N C_N + A_s^o C_s^o + A_P^o C_P^o + B_P^C \tag{5.35}$$

where,

$$A_P = \frac{\Delta y}{\Delta t} + \frac{D_n}{\delta\, y_n} + \frac{1}{\left(\frac{\delta y_s}{D_s} + \frac{\Delta t}{L y_1}\right)} \tag{5.36}$$

$$A_N = \frac{D_n}{\delta y_n} \tag{5.37}$$

$$A_s^o = \frac{1}{\left(\frac{\delta y_s}{D_s} + \frac{\Delta t}{L y_1}\right)} \tag{5.38}$$

$$A_P^o = \frac{\Delta y}{\Delta t} \tag{5.39}$$

$$B_P^C = \mu \Delta y S_P^o \tag{5.40}$$

- ***Upper boundary points***

In this case, the boundary flux that must be replaced by the existing boundary condition (Eq. 5.18) and the linear discretized equation for the boundary points is given by:

$$A_P C_P = A_S C_S + A_n^o C_n^o + A_P^o C_P^o + B_P^C \tag{5.41}$$

where,

$$A_P = \frac{\Delta y}{\Delta t} + \frac{D_s}{\delta\, y_s} + \frac{1}{\left(\frac{\delta y_n}{D_n} + \frac{\Delta t}{L y_2}\right)} \tag{5.42}$$

$$A_S = \frac{D_s}{\delta y_s} \tag{5.43}$$

$$A_n^o = \frac{1}{\left(\frac{\delta y_n}{D_n} + \frac{\Delta t}{L y_2}\right)} \tag{5.44}$$

$$A_P^o = \frac{\Delta y}{\Delta t} \tag{5.45}$$

$$B_P^C = \mu \Delta y S_P^o \tag{5.46}$$

Equations 5.29, 5.35 and 5.41 when applied at each of the control volumes form a system of algebraic linear equations whose solution indicates the concentration of free solute within the solid during the water absorption process.

(b) **Solution for the concentration of entrapped solute**

In order to obtain the equations associated to the concentration of entrapped solute in the solid, Eq. (5.15) is integrated in volume and time. Reorganizing the terms, we obtain the following linear algebraic equation, valid for all control volume of the domain:

$$A_P S_P = A_P^o S_P^o + B_P^S \tag{5.47}$$

where,

$$A_P = \frac{\Delta y}{\Delta t} + \mu \Delta y \tag{5.48}$$

$$A_P^o = \frac{\Delta y}{\Delta t} \tag{5.49}$$

$$B_P^S = \lambda \Delta y C_P \tag{5.50}$$

It is important to notice that the Eq. (5.47) is an explicit equation, depending only of the values of the free solute concentration at each nodal point and time.

The total moisture inside the material in a specific position x and instant t is given by summing the amounts of free and entrapped solute as follows:

$$M = S + C \tag{5.51}$$

Equation (5.23) yields the average moisture content of the solid at any time. This equation in the discretized form is written as:

$$\overline{M} = \frac{1}{2a} \sum_{i=2}^{np-1} M_i \Delta y_i \tag{5.52}$$

where np represents the total number of nodal points considered.

The linear algebraic equation system can be solved iteratively using the Gauss–Seidel method, where is assumed that the numerical solution converges when, from the initial condition, the following criterion is satisfied at each nodal point of the domain, at a given time:

$$\left|C_P^{n+1} - C_P^n\right| \leq 10^{-10} \tag{5.53}$$

where n represents the nth iteration at each instant. To obtain the predicted results, a time step and mesh refinement study was performed. After this process a grid with np = 20 nodal points and a time step $\Delta t = 20$ s was chosen.

(c) **Estimation of the model parameters**

For the computer simulation of the water absorption process using the Langmuir-type modl, it is necessary to use D, λ and μ values consistent with the experimental data. This procedure is described in detail below.

- ***Initial estimate of the probabilities λ and μ***

Considering a one-dimensional approach, $\kappa = \pi^2 D/(2R_2)^2$ and satisfying $2\lambda << \kappa$ and $2\mu << \kappa$, the following approximate solution for determining the average moisture content (Eq. 5.24) is valid [5]:

$$\begin{aligned}\frac{\overline{M}}{M_e} = & \frac{\mu}{\lambda+\mu} e^{-\lambda t}\left[1 - \frac{8}{\pi^2}\sum_{n=1}^{\infty}\frac{e^{-\kappa(2n+1)^2 t}}{(2n+1)^2}\right] \\ & + \frac{\mu}{\lambda+\mu}\left(e^{-\mu t} - e^{-\lambda t}\right) + \left(1 - e^{-\mu t}\right)\end{aligned} \tag{5.54}$$

For long times, when $\kappa \times t >> 1$, Eq. (5.54) is reduced to:

$$\frac{\overline{M}}{M_e} = 1 - \frac{\lambda}{\lambda+\mu} \mathrm{Exp}(-\mu \times t) \tag{5.55}$$

which can be written, as follows:

$$\frac{\overline{M}}{M_e} = 1 - A\mathrm{Exp}(-B \times t) \tag{5.56}$$

where

$$A = \frac{\lambda}{\lambda+\mu} \tag{5.57}$$

$$B = \mu \tag{5.58}$$

With the values of the average moisture content over time obtained experimentally, it is possible to perform a nonlinear regression of Eq. (5.56), to obtain the values of A and B statistical parameters and, with those, determine the values of λ and μ parameters using Eqs. (5.57) and (5.58).

- ***Initial estimate of the probabilities λ and μ***

In this case, considering a one-dimensional approach, for short times ($t \leq 0.7/\kappa$), the following approximate solution for determining the average moisture content (Eq. 5.59) is valid [5]:

$$\overline{M} = \frac{4}{\pi^{3/2}}\left(\frac{\mu}{\lambda + \mu} M_e\right)\sqrt{\kappa \times t} \tag{5.59}$$

or yet,

$$\overline{M} = \frac{4}{\pi^{3/2}}\left(\frac{\mu}{\lambda + \mu} M_e\right)\sqrt{\kappa} \times \sqrt{t} \tag{5.60}$$

Thus, substituting $\kappa = \pi^2 D/(2R_2)^2$ in Eq. (5.60) we get:

$$\overline{M} = \frac{4}{\pi^{3/2}}\left(\frac{\mu}{\lambda + \mu} M_e\right)\sqrt{\frac{\pi^2 D}{(2R_2)^2}} \times \sqrt{t} \tag{5.61}$$

Calculating the derivative of Eq. (5.61) with respect to time and since, for initial times, the behavior of the average moisture content is approximately linear with time, we can write:

$$\frac{d\overline{M}}{d\sqrt{t}} \approx \frac{\overline{M}_2 - \overline{M}_1}{\sqrt{t_2} - \sqrt{t_1}} = \left(\frac{4}{\pi^{3/2}}\right)\left(\frac{\mu}{\lambda + \mu} M_e\right)\left(\frac{\pi}{2R_2}\right) \times \sqrt{D} \tag{5.62}$$

which leads to:

$$D = \pi \times \left(\frac{R_2}{2M_e}\right)^2 \left(\frac{\lambda + \mu}{\mu}\right)^2 \left(\frac{\overline{M}_2 - \overline{M}_1}{\sqrt{t_2} - \sqrt{t_1}}\right)^2 \tag{5.63}$$

where $\overline{M}_1$ and $\overline{M}_2$ are values of the average moisture content obtained experimentally, at times t_1 and t_2, respectively, and M_e is the equilibrium moisture content.

An alternative method to estimate the mass diffusion coefficient is to consider $\lambda = 0$ (Langmuir-type model tending to the Fick´s model) and determine D using the least square error technique to minimize the error between theoretical and experimental data (Eq. 5.64).

It is important to notice that the model parameters D, μ, and λ determined as already mentioned are only an initial estimate and that the mathematical procedures reported above can also be used for a three-dimensional analysis.

Table 5.1 Geometrical parameters used in the simulation

Parameter	Value
L (m)	0.3
a (m)	1.5×10^{-3}

- ***Actual estimate of the parameters D, μ, and λ***

Once the initial estimate of parameters D, μ and λ was made, the real estimate of these parameters can be performed, comparing the least square error between the predicted and experimental data of the average moisture content, and varying these parameters until a minimum error is reached, according to Eq. (5.64).

$$ERQM = \sum_{i=1}^{n} \left[\overline{M}_{predict} - \overline{M}_{\exp erimental}\right]^2 \tag{5.64}$$

where n is the number of experimental points.

5.2.4.4 Langmuir-Type Model Application: 1D Approach

Analytical and Numerical Results

- ***Absorbed Water Kinetics***

The exact and numerical solutions of the governing equations were applied to predict moisture diffusion in polymer composites reinforced with caroá fibers. As already mentioned, the composites studied have width and length greater that its thickness. The results that presented here refer to the case where $Ly_2 = Ly_1 = L$ and $R_2 = a$. It is assumed that water penetrates only in the thickness. direction. Table 5.1 presents the geometric parameters used in the simulation.

For the validation of the model, the predicted result of the average moisture content was compared with analytical results reported by Santos et al. [3] and the experimental data reported by Silva [19] and Nóbrega et al. [20] for polymer composite materials reinforced with Caroá fiber (T = 25 °C). From this comparison, the mass diffusion coefficient and the probabilities μ and λ, of the model were estimated using the least squares error technique, given by the following equation:

$$ERQM = \sum_{i=1}^{n} \left[\overline{M}_{predicted} - \overline{M}_{\exp erimental}\right]^2 \tag{5.54}$$

where n is the number of experimental points. According to Santos et al. [3], an average quadratic error of 0.047467 $kg_{water}/kg_{dry\ solid}$) was obtained. Table 5.2 shows the estimated data.

Table 5.2 Physical parameters used in the simulation

Parameter	Value
D (m^2s^{-1})	7.020×10^{-12}
μ (s^{-1})	1.697×10^{-6}
λ (s^{-1})	0.836×10^{-6}

Table 5.3 Some values of the p and k eigenvalues

n	k_n (m^{-1})	P_n (m^{-1})
1	1049.07	-7.5768×10^{-6}
2	1058.21	-1.29658×10^{-6}
3	3142.29	-6.06761×10^{-6}
4	3171.09	-1.44941×10^{-6}
5	5236.41	-1.67169×10^{-6}

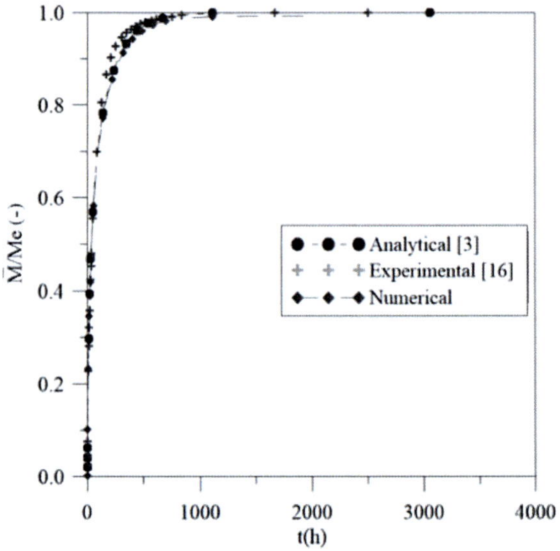

Fig. 5.5 Numerical, analytical and experimental dimensionless average moisture content of polymer composites reinforced with caroá fiber during the water absorption process (T = 25 °C, $Ly_2 = Ly_1 = 0.3$ m)

These values led to 30 pairs of eigenvalues and one pair of imaginary root which were used to obtain the exact solution of Eqs. (5.19), (5.20) and (5.24). Table 5.3 presents five values of p_n and k_n engeinvalues determined by Santos et al. [3].

Figure 5.5 illustrates the comparison between numerical, analytical and experimental data of the average moisture content obtained during water absorption in caroá fiber-reinforced polymer composites immersed in water at 25 °C.

Figure 5.5 shows some discrepancies between experimental and numerical data, which can be attributed to the lack of suitable boundary conditions for the model and the assumption of constant properties. The numerical results, however, showed good

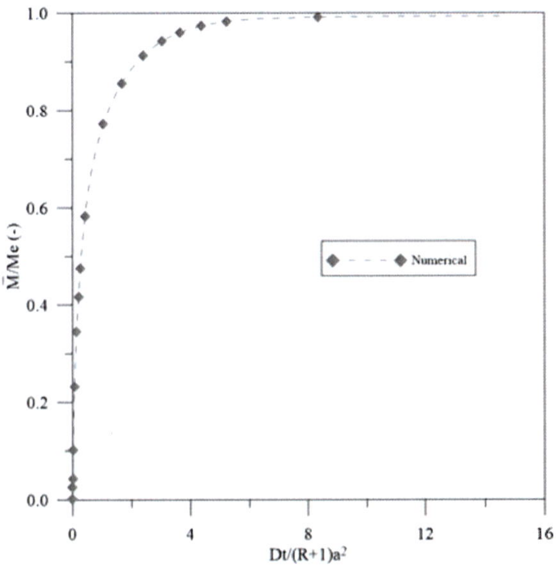

Fig. 5.6 Dimensionless average moisture content of the caroá fiber reinforced polymer composite as a function of modified Fourier number of mass transfer during the water absorption process (T = 25 °C, $Ly_2 = Ly_1 = 0.3$ m)

agreement with analytical results, indicating that the proposed mathematical model properly describes the water diffusion process inside the material.

A more general mathematical analysis was obtained by plotting the dimesionless average moisture content as a function of another dimensioless parameter Fo = $Dt/[(R + 1)a^2]$ namely modified Fourier number for mass transfer. These results are shown in Fig. 5.6.This procedure allows for results to be independent of mass diffusivity, λ and μ probabilities, composite thickness and time. Analysis of this figure indicates that water sorption is very quick in the early stages up to a modified Fourier number of mass transfer of approximately 2.0, and tends to decline for long exposure times until equilibrium is achieved (saturation condition, $M_e = 14.488\%$), where the dimensionless average moisture content tends to a maximum value. The time to achieve the hygroscopic equilibrium was estimated to Fo $\cong 10F_o \approx 12$.

- ***Absorbed Free and Entrapped Water Molecules Distribution***

Figure 5.7 shows the variation of dimensionless free solute concentration (analytical and numerical) along the thickness of the solid, given by C/Ce, where Ce (estimated by Eq. 5.22 as Ce = 0.09778 kg/kg) represents the equilibrium concentration of the solute in the fluid medium. Since the solid in question is symmetrical with respect to its center, results are ploted only from the center (y = 0 m) to the composite surface (y = 0.0015 m). This figure shows that, for shorter times, moisture concentration variation is higher close to the surface of the material, that is, there is a high concentration gradient of free water, in these regions. With increasing time, this relationship tends to approach 1.0, as the hygroscopic equilibrium condition (saturation) is reached. Besides, at any point within the solid, the moisture content increases with time until it reaches equilibrium.

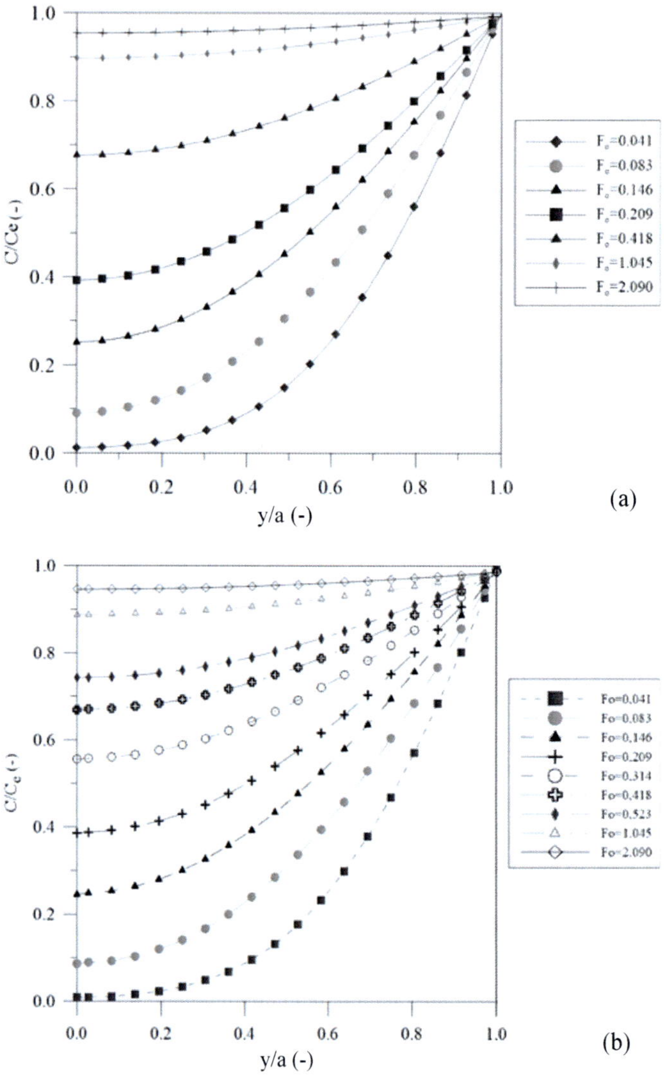

Fig. 5.7 Dimensionless free solute concentration distribution inside the caroá fiber reinforced plymer composite for different modified Fourier number of mass transfer. **a** Analytical and **b** Numerical results (T = 25 °C, Ly2 = Ly1 = 0.3 m)

Figure 5.8 illustrates the distribution of the dimensionless bound (entrapped) water molecules (numerical) along the thickness of the caroá fiber-reinforced polymer composite. Analysis of this figure shows that an increase in the content of bound water molecules depends on the increase in the concentration of free water molecules into the material. The major bound water concentration gradients are found near the solid surface. For longer times, or when there is a greater amount of free water molecules

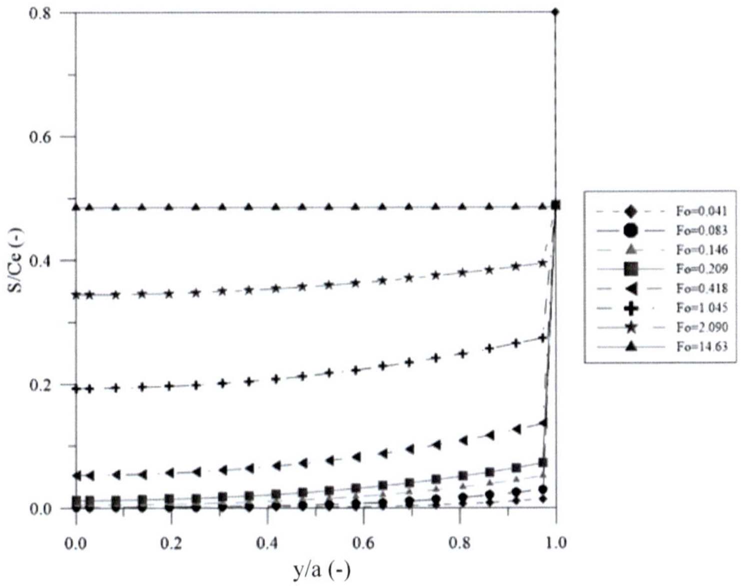

Fig. 5.8 Predicted dimensionless entrapped solute concentration inside the caroá fiber reinforced plymer composite for different modified Fourier number of mass transfer (T = 25 °C, Ly2 = Ly1 = 0.3 m)

within the material, there is a higher number of entrapped water molecules, and this condition occurs until that equilibrium is reached. This occurs as $\partial S/\partial t = 0$, or yet, $\lambda C = \mu S$, which implies that $S/C_e = \lambda/\mu = 0.493$, for $t \to \infty$.

At the beginning, the observed behaviour for water absorption is Fickian, that is, there is moisture migration of water molecules in "free" state. However, as time goes by and as more moisture is absorbed, the water diffusion rate decreases. This behaviour is explained by two phenomena: (a) as water absorbed more molecules are linked to the polymer chains, thus reducing the amount of water that can be absorbed, and (b) the relaxation rate becomes larger than the diffusion rate, controlling the final stages of the process.

Figure 5.9 shows the dimensionless moisture content absorbed into the material, obtained by the sum of the free water molecules concentration and the entrapped water molecules concentration. In regions near the surface, water absorption is faster, because there is a larger area in direct contact with water bath. The water penetrates the interior of the material generating a higher moisture content gradient along the thickness, which decreases with increasing immersion time. Thus, at any point inside the solid, the moisture content increases with time until it reaches equilibrium, i.e., its saturation point. It is intuitive to say that at longer immersion times there is an increase in the amount of molecules entrapped within the material while the amount of free molecules to diffuse decreases.

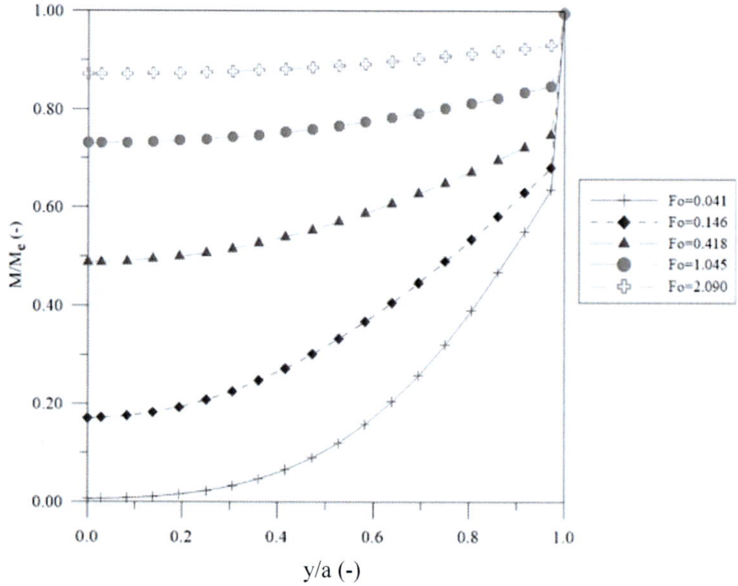

Fig. 5.9 Predicted dimensionless misture content inside the caroá fiber reinforced plymer composite for different modified Fourier number of mass transfer (T = 25 °C, $Ly_2 = Ly_1 = 0.3$ m)

For a more complex physical situation, for example, where $Ly_2 \neq Ly_1$, Melo et al. [21] reported a theoretical analysis aiming to evaluate the effect of the water layer thickness (upper and bottom) on the water absorption behavior inside the material. In that research, the authors considered equilibrium moisture content Me = 0.14488 kg/kg, water bath temperature 25 °C, and total process time 2250 h. Table 1 summarizes the data used in the simulations.

Figure 5.10 illustrates the average moisture content in the material along the process for different arbitrary cases, as reported in Table 5.4. Figure 5.11 shows the local moisture contents obtained from the sum of the free and bound water molecule concentrations along the thickness of the material for different elapsed times and distances from the composite surface to maximum water level in the container. Figure 5.12 shows the behavior of the variation rate of the local moisture content as a function of time in the center of the composite.

According to Melo et al. [21], water absorption kinetics is strongly affected by the water layer thickness, the moisture content gradient, and the equilibrium moisture content inside the material. These authors verified that the higher gradients are found in the regions where Ly_1 or Ly_2 have higher value. Furthermore, the largest concentration gradients of free and entrapped water molecules are at the composite surface, and an asymmetric behavior of the process variables is obtained when the geometric parameters Ly_2 and Ly_1 are different.

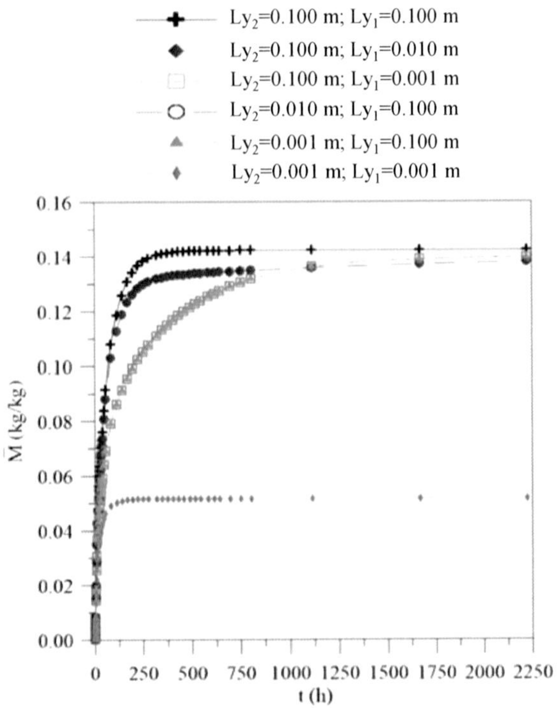

Fig. 5.10 **a** Average moisture content as a function of the process time for different physical saturations. **b** Detailed view of **a**

Table 5.4 Process parameters values of the polymer composite used in the simulations

To (°C)	2a (m)	Ly_2 (m)	Ly_1 (m)	μ (10^{-6} s^{-1})	λ (10^{-6} s^{-1})	D (10^{-12} m^2s^{-1})
25	0.003	0.100	0.100	5	1	5
25	0.003	0.100	0.010	5	1	5
25	0.003	0.100	0.001	5	1	5
25	0.003	0.010	0.100	5	1	5
25	0.003	0.001	0.100	5	1	5
25	0.003	0.001	0.001	5	1	5

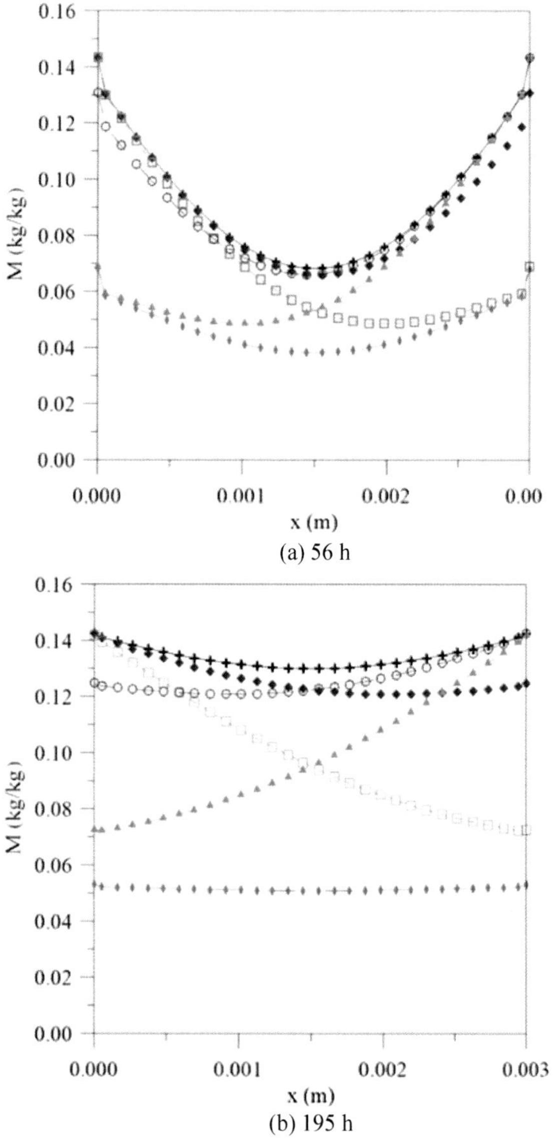

Fig. 5.11 Local moisture content as function of the position inside the material at different process times

5.2.5 *The Water Absorption Process: 3D Approach*

5.2.5.1 Solution Techniques: 3D Approach

Numerical Solution

The finite volume method was used for the three-dimensional numerical solution of the governing equations applied to moisture absorption in polymer composites

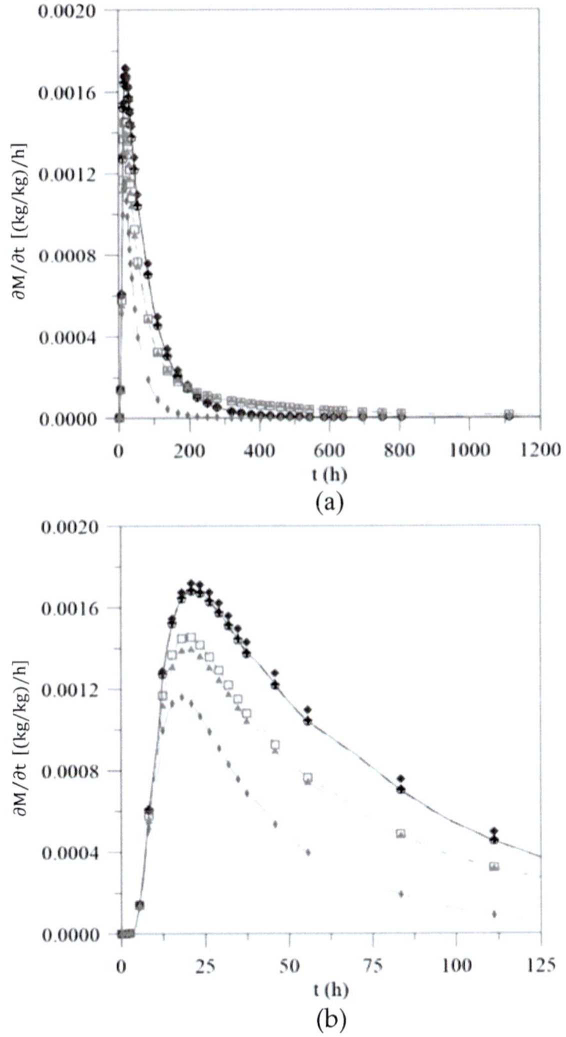

Fig. 5.12 Temporal variation rate of the local moisture content in the center of the polymer composite **a** and detail of the figure **b**

with parallelepipedic form, as reported by Santos et al. [22, 23], Brito et al. [24] and Santos [25]. In this solution, a fully implicit formulation for the concentration of free solute and explicit formulation for the concentration of trapped solute were used. Figure 5.13 illustrates a control volume (sub domain) used for discretization of the governing equations. The nodal point P (in the center of the control volume), its adjacent neighbors W, E, S, N, T, and F, the distances between these nodal points, and the dimensions Δx, Δy, and Δz, of the control volume are also shown in this Figure.

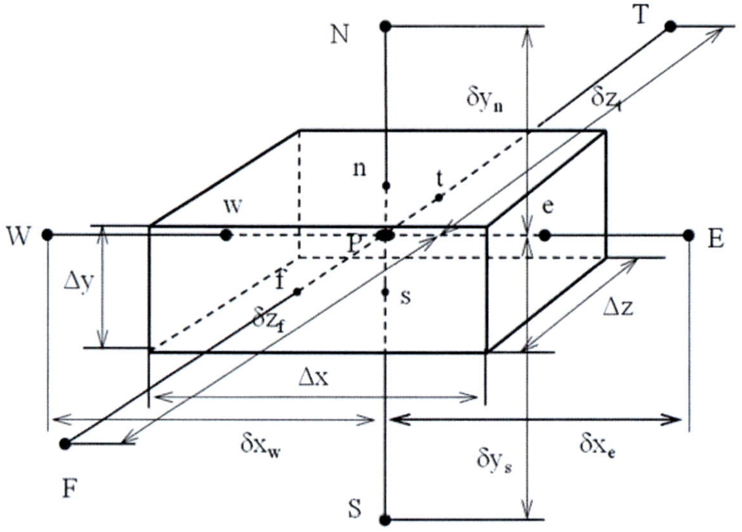

Fig. 5.13 Three-dimensional control volume used in this work

Complete detailing of the numerical procedure used for the solution of the governing equations is given below.

(a) **Solution for the Free Water Molecules Concentration**

- ***Internal points***

The numerical solution of Eq. (5.3) is obtained by integrating it in volume and time. Assuming a fully implicit formulation, one can write the Eq. (5.3) in its discretized linear form as follows:

$$A_P C_P = A_E C_E + A_W C_W + A_N C_N + A_S C_S + A_F C_F + A_T C_T$$

$$+ A_P^o C_P^o + B_P^C \tag{5.66}$$

where:

$$A_E = \frac{D_e}{\delta x_e} \Delta y \Delta z \tag{5.67}$$

$$A_W = \frac{D_w}{\delta x_w} \Delta y \Delta z \tag{5.68}$$

$$A_N = \frac{D_n}{\delta y_n} \Delta x \Delta z \tag{5.69}$$

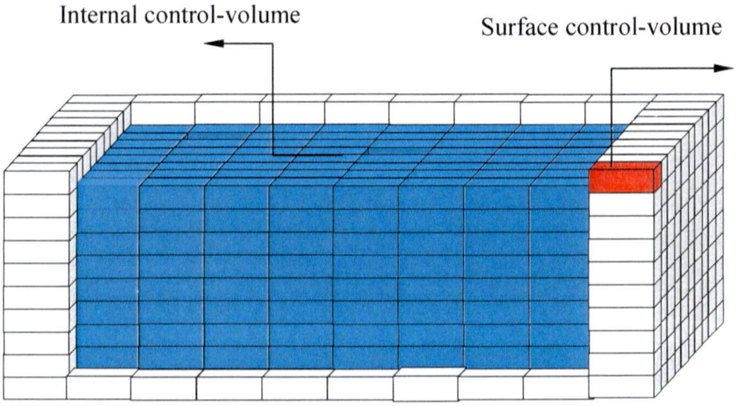

Fig. 5.14 Numerical grid showing the internal and surface control-volumes

$$A_S = \frac{D_s}{\delta y_s} \Delta x \Delta z \tag{5.70}$$

$$A_F = \frac{D_f}{\delta z_f} \Delta y \Delta x \tag{5.71}$$

$$A_T = \frac{D_t}{\delta z_t} \Delta y \Delta x \tag{5.72}$$

$$A_P^o = \frac{\Delta x \Delta y \Delta z}{\Delta t} \tag{5.73}$$

$$A_P = \frac{\Delta x \Delta y \Delta z}{\Delta t} + \frac{D_e}{\delta x_e} \Delta y \Delta z + \frac{D_w}{\delta x_w} \Delta y \Delta z + \frac{D_n}{\delta y_n} \Delta x \Delta z + \frac{D_s}{\delta y_s} \Delta x \Delta z$$

$$+ \frac{D_f}{\delta z_f} \Delta y \Delta x + \frac{D_t}{\delta z_t} \Delta y \Delta x \tag{5.74}$$

$$B_P^C = -\left(S_P - S_P^o\right) \frac{\Delta x \Delta y \Delta z}{\Delta t} \tag{5.75}$$

- ***Boundary points***

It should be noted that Eq. (5.66) is only applied to the internal control volumes of the computational domain (Fig. 5.14). For the other control volumes (symmetry and border), a mass balance in each one of them is performed. In total, there are 27 different types of control volumes. As an example, the result of Eq. 5.3 applied to the control volume of the right upper corner of the computational domain, as shown in Fig. 5.14, is given by the following equation:

$$A_P C_P = A_W C_W + A_S C_S + A_T C_T + A_P^o C_P^o + B_P^C \tag{5.76}$$

where:

$$A_W = \frac{D_w^C}{\delta x_w} \Delta y \Delta z \tag{5.77}$$

$$A_S = \frac{D_s}{\delta y_s} \Delta x \Delta z \tag{5.78}$$

$$A_T = \frac{D_t}{\delta z_t} \Delta y \Delta x \tag{5.79}$$

$$A_P^o = \frac{\Delta x \Delta y \Delta z}{\Delta t} \tag{5.80}$$

$$A_P = \frac{\Delta x \Delta y \Delta z}{\Delta t} + \frac{D_w}{\delta x_w} \Delta y \Delta z + \frac{D_s}{\delta y_s} \Delta x \Delta z + \frac{D_t}{\delta z_t} \Delta y \Delta x + \frac{\Delta y \Delta x}{\left(\frac{\delta z_f}{D_f} + \frac{\Delta t}{l_z}\right)} + \frac{\Delta z \Delta x}{\left(\frac{\delta y_n}{D_n} + \frac{\Delta t}{l_y}\right)} + \frac{\Delta y \Delta z}{\left(\frac{\delta x_e}{D_e} + \frac{\Delta t}{l_x}\right)} \tag{5.81}$$

$$B_P^C = \frac{C_f^o \Delta x \Delta y}{\left(\frac{\delta z_f}{D_f} + \frac{\Delta t}{l_z}\right)} + \frac{C_e^o \Delta y \Delta z}{\left(\frac{\delta x_e}{D_e} + \frac{\Delta t}{l_x}\right)} + \frac{C_n^o \Delta x \Delta z}{\left(\frac{\delta y_n}{D_n} + \frac{\Delta t}{l_y}\right)} - \left(S_P - S_P^o\right) \frac{\Delta x \Delta y \Delta z}{\Delta t} \tag{5.82}$$

(b) **Solution for the entrapped water molecules concentration**

The numerical solution of Eq. (5.4) is obtained by integrating it in volume and time. Assuming an explicit formulation, one can write the Eq. 5.4, in its discretized linear form, as follows:

$$A_P S_P = A_P^o S_P^o + B_P^S \tag{5.83}$$

where:

$$A_P^o = \frac{\Delta x \Delta y \Delta z}{\Delta t} \tag{5.84}$$

$$A_P = \frac{\Delta x \Delta y \Delta z}{\Delta t} + \mu \Delta x \Delta y \Delta z \tag{5.85}$$

$$B_P^S = \lambda C_P \Delta x \Delta y \Delta z \tag{5.86}$$

In the discretized form, the local and average moisture contents can be written, respectively, as follows:

$$M_{i,j,k} = C_{i,j,k} + S_{i,j,k} \tag{5.87}$$

$$\overline{M} = \frac{1}{V} \sum_{i=2}^{npx-1} \sum_{j=2}^{npy-1} \sum_{k=2}^{npz-1} M_{i,j,k} \Delta V_{i,j,k} \tag{5.88}$$

in which i, j e k represent the position of the nodal point in the x, y, and z directions, respectively, and npx, npy and npz are the nodal point numbers in the x, y, and z directions respectively.

From the discretization of the governing equations, a system of liear algebraic equations is generated that must be solved to obtain the values of C, S and M within the material throughout the process. The Gauss–Seidel iterative method can be used for solving this system of algebraic equations. In order to obtain the numerical results, simulations were performed using a grid with $20 \times 20 \times 20$ nodal points and time step $\Delta t = 20$ s. Other details about this numerical procedure can found in the references already cited in this chapter.

5.2.5.2 Langmuir-Type Model Application: Three-Dimensional Approach

Numerical Results

Based on the references already mentioned, the influence of the geometrical parameter L_y (see Fig. 5.2), in the process of water absorption in fiber-reinforced polymer composites was evaluated. For this analysis, the authors considered $\mu = \lambda = 1.0 \times 10^{-6}\ s^{-1}$, $D = 1 \times 10^{-12}\ m^2/s$, $R_1 = R_3 = 0.0100$ m, $R_2 = 0.0015$ m, $R_3 = 0.000$ m, $Lx_1 = Lx_2 = 0.1000$ m, $Lz_1 = Lz_2 = 0.1000$ m and $Ly_1 = Ly_2 = 0.0010$ m, 0.0100 m and 01,000 m,

The obtained results indicated that variations in L_y values strongly affect the average moisture content, and also average free and entrapped water molecules concentrations. The higher the L_y value, the higher the water layer close to the composite wall and the faster the water absorption rate. For initial times of the process, the differences among the predicted results are not significant, however, in the course of the process a different behavior of the water absorption kinetics can be clearly observed.

Upon analyzing the free water molecules concentration distribution was verified that, for small value of L_y, there is essentially no free water flux in the y-direction but, in the opposite directions, the water flux is more intense and free water molecules move horizontally (x—and z—directions) with nearly equal velocities.

A similar behavior for the distribution of entrapped water molecules concentration to that presented by free solute concentration was observed. In general, the geometric parameter L_y affects both the distribution of free and entrapped solute concentration and, as a result, the total moisture content of the composite immersed in water.

Besides, it was observed small values of the bound water molecules concentration when compared to unbound water molecules concentration. Furthermore, a transient analysis of the predicted results showed that the free water molecules concentration is increasing from the composite surface to its center with a horizontal flux of moisture. Compared with the y-direction, the greatest moisture fluxes occur in x and z- direction. The same behavior is observed when the entrapped water molecules concentration is considered. Nevertheless, entrapped water concentration increases more slowly than free water molecules concentration, mainly at the initial process times.

As final comment, we notice that water absorption is facilitated when polymer molecules have clusters capable of forming hydrogen bonds. Plant fibers are rich in cellulose, hemicellulose and lignin which have hydroxyl groups, i.e., have high affinity for water. The absorption of water by the resin, in turn, can be considered practically null, since it has a considerable hydrophobic character [26]. Addition of the plant fibers to hydrophobic resins leads to an increase in the water absorption levels, so an important parameter to be analyzed is how much water is absorbed by the material over time.

The effects caused by long time exposure to moisture may be irreversible due to the water molecules affinity with specific functional groups of the polymeric matrices. Destructive changes usually occur due to degradation of existing physical–chemical interactions between the resin and fiber and, as a consequence, there are changes in the fiber, causing delamination and reduction in the composite material properties. Thus, understanding of the water absorption process is crucial to predict the quality of the material e wet environments.

From the physical and mathematical point of views is important to analyse quantitatively what happens with the general solution for the extreme values of the probability μ which correspond to very fast or very slow process. When μ is very large as compared to λ, the process is very rapid compared with diffusion. In this situation, the immobilized component is in equilibrium with the component free to diffuse into the composite, or yet, the number of free molecules that become entrapped and remain entrapped by a time long enough to hinder diffusion is very small, and process is controlled by diffusion. However, if $\mu \rightarrow 0$, the process is infinitely slow, the composite is occuped, by simple diffusion, and only a fraction of solute (moisture) can be accommodated in the freely diffusing state and none in the immobilized state. On the other hand, when λ increases sufficiently, moisture absorption is hindered due to the higher probability that a free water molecule will become entrapped instead of freely diffusing into the composite [10, 13]. Further, for the case where D is very large, the diffusion is so rapid that the concentration of free and immobilized solute (water) are almost uniform through the composite during the water uptake process.

From the explanation above it is possible to conclude that variations in the mass diffusion coefficient and the probabilities that water molecules are free or entrapped, deeply modifies the water absorption process, which proves the great potential of the Langmuir-type model and its ability of adequately describing moisture migration behaviour inside the polymer composite.

References

1. Bonniau, P., Bunsell, A.R.: A comparative study of water absorption theories applied to glass epoxy composites. J. Compos. Mater. **15**(3), 272–293 (1981)
2. Grace, L.R., Altan, M.C.: Characterization of anisotropic moisture absorption in polymeric composites using hindered diffusion model. Compos. A Appl. Sci. Manuf. **43**(8), 1187–1196 (2012)
3. Santos, W.R.G., Melo, R.Q.C., Lima, A.G.B.: Water absorption in polymer composites reinforced with vegetable fiber using Langmuir-type model: An exact mathematical treatment. Defect Diffus. For. **371**, 102–110 (2016)
4. Perreux, D., Suri, C.: A study of the coupling between the phenomena of water absorption and damage in glass/epoxy composite pipes. Compos. Sci. Technol. **57**(9–10), 1403–1413 (1997)
5. Carter, H.G., Kibler, K.G.: Langmuir-type model for anomalous moisture diffusion in composite resins. J. Compos. Mater. **12**(2), 118–131 (1978)
6. Glaskova, T.I., Guedes, R.M., Morais, J.J., Aniskevich, A.N.: A comparative analysis of moisture transport models as applied to an epoxy binder. Mech. Compos. Mater. **43**(4), 377–388 (2007)
7. Cotinaud, M., Bonniau, P., Bunsell, A.R.: The effect of water absorption on the electrical properties of glass-fibre reinforced epoxy composites. J. Mater. Sci. **17**(3), 867–877 (1982)
8. Apicella, A., Estiziano, L., Nicolais, L., Tucci, V.: Environmental degradation of the electrical and thermal properties of organic insulating materials. J. Mater. Sci. **23**(2), 729–735 (1988)
9. Popineau, S., Rondeau-Mouro, C., Sulpice-Gaillet, C., Shanahan, M.E.: Free/bound water absorption in an epoxy adhesive. Polymer **46**(24), 10733–10740 (2005)
10. Crank, J.: The mathematics of diffusion, 2nd edn. Oxford University Press, Oxford (1975)
11. Fu, Z., Chen, W., Yang, H.: Boundary particle method for laplace transformed time fractional diffusion equations. J. Comput. Phys. **235**, 52–66 (2013)
12. Zhu, S., Satravaha, P., Lu, X.: Solving linear diffusion equations with the dual reciprocity method in Laplace space. Eng. Anal. Bound. Elem. **13**(1), 1–10 (1994)
13. Grace, L.: Non-Fickian three-dimensional moisture absorption in polymeric composites: development and validation of hindered diffusion model. Ph.D Thesis, University of Oklahoma, Norman, USA, (2012)
14. Luikov, A.V.: Analytical heat diffusion theory, p. 684. Academic Press, Inc. Ltd., London (1968)
15. Carslaw, H.S., Jaeger, J.C.: Conduction of heat in solids. 2nd edn, p. 510. University Press, Oxford, New York (1959)
16. Melo, R.Q.C., Santos, W.R.G., Lima, A.G.B.: Applying the Lagmuir-type model on the water absorption in vegetable fiber reinforced polymer composites: A finite-volume approach. In: XXXVIII Iberian Latin-American Congress on Computational Methods in Engineering, Florianópolis, Brazil. (2017)
17. Patankar, S.V.: Numerical heat transfer and fluid flow. Hemisphere Publishing Corporation, New York (1980)
18. Maliska, C.R.: Computational heat transfer and fluid mechanics. LTC, Rio de Janeiro (2004).(In Portuguese)
19. Silva, C.J.: Water absorption in composite materials of vegetal fiber: modeling and simulation via CFX. Master Dissertation in Mechanical Engineering, Federal University of Campina Grande, Brazil (2014) (In Portuguese)
20. Nóbrega, M.M.S., Cavalcanti, W.S., Carvalho, L.H., Lima, A.G.B.: Water absorption in unsaturated polyester composites reinforced with caroá fiber fabrics: modeling and simulation. Mat. -wiss. u.Werkstofftech. 41(5), 300–305 (2010)
21. Melo, R.Q.C., Fook, M.V.L., Lima, A.G.B.: Non-fickian moisture transport in vegetable-fiber-reinforced polymer composites using a Langmuir-type model. Polymers 12, 2503 (2020)
22. Santos, W.R.G., Melo, R.Q.C., Correia, B.R.B., Magalhães, H.L.F., Cabral, E.M., Figueiredo, M.J., Lima, A.G.B.: Water absorption in vegetable fiber-reinforced polymer composites: A

three-dimensional investigation using the Langmuir-type model. Defect Diffus For **399**, 164–170 (2020)
23. Santos, W.R.G., Brito, M.K.T., Lima, A.G.B.: Study of the moisture absorption in polymer composites reinforced with vegetal fiber using Langmuir's model. Mater. Res. 22, e20180848 (2019)
24. Brito, M.K.T., Santos, W.R.G., Correia, B.R.B., Queiroz, R.A., Tavares, F.V.S., Oliveira Neto, G.L., Lima, A.G.B.: Moisture absorption in polymer composites reinforced with vegetable fiber: a three-dimensional investigation via Langmuir model. Polymers 11, 1847 (2019)
25. Santos, W.R.G.: Heat and mass transfer in polymeric composites reinforced by vegetable fibers: advanced modeling and simulation. Campina Grande: Doctoral Thesis in Process Engineering, Federal University of Campina Grande, Campina Grande, Brazil, (2019)
26. Sanchez, E.M., Cavani, C.S., Leal, C.V., Sanchez, C.G.: Composites of unsaturated polyester resin with sugarcane bagasse: influence of fiber treatment on properties. Polymer **20**(3), 194–200 (2010)

Chapter 6
Conclusions

This book is entirely dedicated to diffusion processes in polymer composites, with particular reference to moisture absorption in vegetable fiber-reinforced polymer composites. An overview of basic issues related to this class of physical problem including fundamentals, experiments, advanced mathematical modeling, and engineering applications are shown in detail. Special attention is given to unsaturated polyester composites reinforced by caroá and macambira vegetable fibers.

Composite is a multiphase material consisting of a continuous phase (matrix) and one or more dispersed phases (reinforcement). The majority of the composites use thermoset matrices, with particular attention to epoxies and polyesters. Vegetable fibers may be used reinforment in polymer composites.

The use of raw materials from renewable sources, such as, vegetable fibers have been the subject of several sudies for many years due to different reasons such as high potential for diverse industrial and technological applications especially those related to socio-economic and environmental benefits.

Despite of their importance, environmental factors such as, moisture and temperature, strongly affect the durability, mechanical properties and many other characteristics of vegetable fiber-reinforced polymer composites. The hydrophilic nature of the vegetable fibers leads to weak interfacial adhesion which causes a reduction in the mechanical properties of the composite. Thus, it can be stated that these materials are very sensitive to water and heat and, to address new researches related to water absorption process in vegetable fiber-reinforced polymer composites, it is crucial to control the process in order to obtain good quality composites when in operation.

The contents of this book evince that the mechanisms governing water absorption in polymer composite are very different from those which are encountered in traditional materials, and requires a full understanding of their constituents (reinforcing and matrix) and their interaction at the fiber-matrix interface (adhesion) or interphase. The adhesion quality is vital to obtain high mechanical performance of the composite, avoiding failure and premature damage in the composite (matrix cracking, local deterioration, delamination, fiber debonding, and pull-out). Thus, the present book deals specifically with water absorption in vegetable fiber-reinforced

J. M. P. Q. Delgado et al., *Moisture Transport in Polymer Composite Materials*, Advanced Structured Materials 160, https://doi.org/10.1007/978-3-030-77826-2_6

unsaturated polyester composites. Emphasis is given to unsaturated polyester matrix composites reinforced with caroá, macambira and sisal natural fibers by theoretical (analytical and numerical) and experimental techniques.

Herein, based on the Fick´s second law and non-Fickian diffusion model (Langmuir´s model) different transient macroscopic models were proposed, and the mathematical formalism based on the finite-volume method (numerical solution), Laplace transform and Separation of variables (exact solution) were used to solve the governing equations. Experimental tests for water absorption were performed by immersing the composite samples in a bath of distilled water at different temperatures, and the water uptake was measured gravimetrically along the process. The unsaturated polyester composites reinforced by caroá, macambira and sisal fibers were manufactured by hand lay-up. For both, the theoretical and experimental analyses, the effect of the composite dimensions, water bath temperature, and water layer thickness above the composite surface were evaluated. Microscopic images on the vegetable fibers were also analyzed.

The content of this book has demonstrated that the water uptake process and its effects in fiber-reinforced polymer composites is now a well-understood topic.

From the experimental and theoretical results of apparent weight gain of the polymer composites associated with water absorption, it can be concluded that mechanical properties (Tensile strength, Young's modulus, impact resistance, and elongation at break) of these materials effectively are affected by moisture content and temperature. The proposed mathematical models adequately describe the process of water diffusion inside the fiber-reinforced polymer composites and simulations proved to be an essential tool in understanding the process. The Langmuir-type model proved to be able to predict both Fickian and non-Fickian moisture absorption with equal ease. For this purpose, it is only necessary to change the problabilities parameters intrinsic to the model to make it truly comprehensive.

It was verified that water absorption by composites increases with fiber loading and is higher than for the polymer matrix alone. Temperature strongly influences the kinetics of water absorption and the absorption rate is higher at higher temperatures. Also, the highest rates of water diffusion were obtained in the first stages of sorption, and the sorption rate decreases at longer water immersion times. At lower temperatures, sample thickness affects water absorption more than temperature, but, at higher temperatures, the opposite occurs, i.e., the temperature in the water absorption were more relevant than that due to changes in the sample´s area/volume ratio.

The knowledge of moisture distribution inside the polymer composite is very important and allows identification of more propicious areas for delamination problems (moisture induced degradation) due to the weakness of the fiber-matrix interface and, consequently, reduction in the composites mechanical properties. The data presented demonstrates that the regions in the neighborhood of the vertices (corners) of the polymer composite (rectangular prism) had the highest mass transfer rates, being more susceptible to crack and deformation.

The demands of vegetable fibers has effectively increased in the composite industry. Reasons for this increased demands include their ecofriendly nature, light weight, good set of mechanical properties, worldwide abundance and availability and low cost which effectively reduce the cost of composite materials and increase their sustainability. Vegetable fiber reinforced polymer composites have some disadvantages already mentioned in this book, particularly their sensitivity to moisture and temperature, which limit their widespread applications and affect service life. Future trends in polymer composite materials reinforced by vegetable fibers must focus on guaranteeing property control, so that there are minimum variations in these properties. There is a need to investigate ways to increase their mechanical propertries, thermal stability and decrease their moisture sensitivity through processing, drying and/or chemical modifications. Studies to define universal testing standards and establish a worldwide database directed to numerical and experimental investigation of continuous intrinsic problems are needed. These must include shape, dimension, manufacturing technique, composition, and microstructure of the samples, moisture absorption and other related effects, thermal degradation, inadequate toughness, and reduced long-term stability for outdoor application that require high load or yet high working temperatures. Therefore, research to improve the performance of vegetable fiber reinforced polymer composites such as thermal and chemical treatments, use of compatibilizer, addition of other fillers, polymer coating, and filler hybridization are strongly recommended. As for the matrix, the development of biodegradable polymers and blends, and optimization in the use of recyclable polymers are in demand. Under the mathematical point of view, the development of robust mathematical modelis to predict mechanical behavior of these polymer composites and related effects, especially in moist, corrosive, saline and heated environments, are also recommended.

Finally, the expectation of the authors is that the information outlined in this book may help researchers, designers, engineers and academics in their studies and making design decisions on the use of complex systems such as polymer composites, especially those reinforced with vegetable fiber, in technological applications.

o/k

This sub-series of the SEMA SIMAI Springer Series aims to publish some of the most relevant results presented at the ICIAM 2019 conference held in Valencia in July 2019.

The sub-series is managed by an independent Editorial Board, and will include peer-reviewed content only, including the Invited Speakers volume as well as books resulting from mini-symposia and collateral workshops.

The series is aimed at providing useful reference material to academic and researchers at an international level.

More information about this subseries at http://www.springer.com/series/16499

Francesc Font • Tim G. Myers
Editors

Multidisciplinary Mathematical Modelling

Applications of Mathematics to the Real World

Springer

Editors
Francesc Font
Centre de Recerca Matemàtica
Bellaterra
Barcelona, Spain

Tim G. Myers
Centre de Recerca Matemàtica
Bellaterra
Barcelona, Spain

ISSN 2199-3041 ISSN 2199-305X (electronic)
SEMA SIMAI Springer Series
ISSN 2662-7183 ISSN 2662-7191 (electronic)
ICIAM 2019 SEMA SIMAI Springer Series
ISBN 978-3-030-64271-6 ISBN 978-3-030-64272-3 (eBook)
https://doi.org/10.1007/978-3-030-64272-3

This Springer imprint is published by the registered company Springer Nature Switzerland AG.
The registered company address is: Gewerbestrasse 11, 6330 Cham, Switzerland

Preface

Since 2014, the Centre de Recerca Matemática (CRM) received PhD and post-doctoral funding from the La Caixa Foundation and the CERCA Programme of the Generalitat de Catalunya to carry out "interdisciplinary and collaborative research". These projects have been supervised by members of the CRM and also university researchers from biology, physics, nanoscience as well as local research centres. This has led to a broad range of research lines where, due to the influence of the non-mathematical partners, the focus has been firmly on practical problems.

In this book, we present work from a selection of talks resulting from this research and presented at the "International Congress on Industrial and Applied Mathematics", held in Valencia, 2019. The various chapters describe a wide variety of topics: cancer modelling, carbon capture by adsorption, nanoscale diffusion and complex systems to predict earthquakes. These mathematical studies were specifically aided via collaborations with biomedical engineers, physicists and chemists.

The wide range of topics described in this book reflects not only the multidisciplinary nature of the La Caixa programme but also the true versatility of mathematics.

Barcelona, Spain
August 2020

Francesc Font
Tim G. Myers

Contents

Spatiotemporal Dynamics of Cancer Phenotypic Quasispecies Under Targeted Therapy 1
Celia Penella, Tomás Alarcón, and Josep Sardanyés

Guyer–Krumhansl Heat Conduction in Thermoreflectance Experiments 21
Matthew G. Hennessy and Tim G. Myers

A Mathematical Model of Carbon Capture by Adsorption 35
Francesc Font, Tim G. Myers, and Matthew G. Hennessy

Diffusion Processes at Nanoscale 49
Claudia Fanelli

Maximum Likelihood Estimation of Power-Law Exponents for Testing Universality in Complex Systems 65
Víctor Navas-Portella, Álvaro González, Isabel Serra, Eduard Vives, and Álvaro Corral

Editors and Contributors

About the Editors

Francesc Font received his PhD in Applied Mathematics from the Universitat Politècnica de Catalunya in 2014, and he is currently a Juan de la Cierva research fellow at the Centre de Recerca Matemàtica. He specialises in the mathematical modelling of transport phenomena and has worked on topics such as nanoscale heat transfer, Li-ion batteries or cell motility. Part of his research is in collaboration with industry, and he has been involved in industrial mathematics workshops throughout Europe.

Tim G. Myers is a Senior Researcher at the Centre de Recerca Matemàtica, Adjunct Professor at the Universitat Politecnica de Catalunya and Adjunct Professor of Industrial Mathematics at the University of Limerick. He has been involved in a wide variety of Industrial Mathematics initiatives and is currently the co-ordinator for all European Study Groups with Industry. He is a co-author of "Optics Near Surfaces and at the Nanometer Scale" and has written over 90 journal publications on a range of applied mathematics topics.

Contributors

Tomás Alarcón ICREA, Barcelona, Spain
Departament de Matemàtiques, Universitat Autònoma de Barcelona, Barcelona, Spain
Centre de Recerca Matemàtica, Bellaterra, Barcelona, Spain

Álvaro Corral Barcelona Graduate School of Mathematics, Barcelona, Spain
Complexity Science Hub Vienna, Vienna, Austria
Departament de Matemàtiques, Universitat Autònoma de Barcelona, Barcelona, Spain
Centre de Recerca Matemàtica, Bellaterra, Barcelona, Spain

Claudia Fanelli Universitat Politècnica de Catalunya, Barcelona, Spain

Francesc Font Centre de Recerca Matemàtica, Bellaterra, Barcelona, Spain

Álvaro González Centre de Recerca Matemàtica, Bellaterra, Barcelona, Spain
GFZ German Research Centre for Geosciences, Potsdam, Germany

Matthew G. Hennessy Mathematical Institute, University of Oxford, Oxford, UK

Tim G. Myers Centre de Recerca Matemàtica, Bellaterra, Barcelona, Spain

Víctor Navas-Portella Centre de Recerca Matemàtica, Bellaterra, Barcelona, Spain
Barcelona Graduate School of Mathematics, Barcelona, Spain
Facultat de Matemàtiques i Informàtica, Universitat de Barcelona, Barcelona, Spain

Celia Penella Universitat Pompeu Fabra, Barcelona, Spain

Josep Sardanyés Centre de Recerca Matemàtica, Bellaterra, Barcelona, Spain

Isabel Serra Centre de Recerca Matemàtica, Bellaterra, Barcelona, Spain
Barcelona Supercomputing Center, Barcelona, Spain

Eduard Vives Departament de Matèria Condensada, Facultad de Física, Universitat de Barcelona, Barcelona, Catalonia, Spain
Universitat de Barcelona Institute of Complex Systems (UBICS), Facultat de Física, Universitat de Barcelona, Barcelona, Catalonia, Spain

Spatiotemporal Dynamics of Cancer Phenotypic Quasispecies Under Targeted Therapy

Celia Penella, Tomás Alarcón, and Josep Sardanyés

Abstract Cancer cells have an enormous genetic and phenotypic heterogeneity. Despite modelling this heterogeneity is not trivial, several mathematical and computational models have used the so-called quasispecies theory. This theory, originally conceived to describe the evolution of information in prebiotic systems, has also been applied to investigate fast evolving replicons with large mutation rates, such as RNA viruses and cancer cells. Here, we investigate a quasispecies system composed of healthy and cancer cells with different phenotypic traits. The phenotypes of tumour cells are coded by binary strings including three different compartments with genes involved in cells' proliferation, in genomic stability, and the so-called housekeeping genes. Previous works have studied this system in well-mixed settings with autonomous ordinary differential equations and stochastic bit-string models. Here, we extend the stochastic bit-strings approach to a spatially explicit system using a cellular automaton (CA). In agreement with the prediction of the well-mixed systems, the spatial one also shows a transition towards tumour extinction at increasing tumour cells' mutation rates, displaying however different stationary distributions of cancer phenotypes. We also use the CA to simulate targeted cancer therapies against different tumour phenotypes. Our results indicate that a combination therapy targetting the fastest proliferative cancer cells with and without

C. Penella
Universitat Pompeu Fabra, Barcelona, Spain

T. Alarcón
ICREA, Barcelona, Spain

Departament de Matemàtiques, Universitat Autònoma de Barcelona, Cerdanyola del Vallès, Barcelona, Spain

Centre de Recerca Matemàtica, Bellaterra, Barcelona, Spain
e-mail: talarcon@crm.cat

J. Sardanyés (✉)
Centre de Recerca Matemàtica, Bellaterra, Barcelona, Spain
e-mail: jsardanyes@crm.cat

F. Font, T. G. Myers (eds.), *Multidisciplinary Mathematical Modelling*,
SEMA SIMAI Springer Series 11, https://doi.org/10.1007/978-3-030-64272-3_1

anomalies in the genomic stability compartment is the most effective therapy. Also, a single target of fastest replicative phenotypes is much more effective than targeting cancer cells with anomalies only in the genomic stability compartment.

1 Introduction

Cancer is a genetic disease caused by alterations in the DNA of cells resulting in malignant cell populations that share certain characteristics, such as chronic proliferation, high genomic instability, or resistance to programmed cell death, among others. These common traits are known as the hallmarks of cancer, and although there is a broad diversity of tumour cells both at the intratumour [1] and intertumour [2] levels, cancer cells have common physiological alterations and dynamical properties [3, 4]. Sources of genotypic and phenotypic variability may be multiple (see the review [5] and references therein). For instance, epigenetic mechanisms and other factors such as the tumour microenvironment and genomic instability seem to be a key factor for the increase of mutation rates and genomic anomalies during the life cycle of cancer cells [4] (by means of the so-called mutator phenotype [6–8]). Genomic instability and increased proliferation rates have been suggested to play a key role in the emergence of highly heterogeneous and diverse populations of cancer cells. This is especially important in most advanced tumours [9, 10]. This lack of clonal structure of cancer cells can favour the overcoming of cell proliferation checkpoints or selection barriers such as the action of the immune system or anti-tumour therapies [3, 11].

The mutation rate of normal cells is about 1.4×10^{-10} changes per nucleotide and replication cycle. On the contrary, tumour cells are known to accumulate many mutations and other genetic anomalies such as chromosomal re-arrangements or breaks. The spontaneous mutation rate of cells may not be sufficient to account for this large number of genetic changes and alterations [6, 7, 12–15]. Indeed, research on mutation frequencies in microbial populations with inactivated mismatch repair genes resulted in 10^2–10^3 times the background mutation rate, with comparable increases in cancer cells [16, 17]. Previous results suggest that cancer cells may have an upper limit to the accumulation of mutations and genetic anomalies, being able to achieve a kind of error threshold [18–20]. The error threshold has been widely investigated in the origins of life problem [21, 22] and virology [23, 24]. The quasispecies concept refers to a highly heterogeneous population of genetically related genomes that remain at a mutation–selection equilibrium balance [21–23]. Typically, the quasispecies model describes the dynamics of different sequences (e.g., genomes) under the processes of replication, mutation, and selection [21, 22].

More recently, the quasispecies concept has been extended at the phenotypic level, considering that the sequences, more than being genomes, are compartments or parts of the genome responsible for determining phenotypic traits. This approach has been used to investigate the impact of mutational fitness effects on virus populations [25] as well as to study the dynamics of competition between healthy and tumour cells with heterogeneous populations (e.g., differential proliferation rates, increased mutation rates, etc), giving place to different phenotypes [26–28]. Specifically, Refs. [26, 27] employed differential equations to model this minimal system allowing for phenotypic variability (see Fig. 1),

where a population of healthy cells competes with a population of tumour cells with different phenotypes, which can arise due to mutations or to genetic anomalies. These phenotypes are determined by a 3-bit sequence including different compartments with genes responsible for cell's proliferation, cell's genomic instability, and house-keeping genes. By means of replication and mutation, tumour cells can produce a heterogeneous population composed of different phenotypes (see Sect. 2 for further details and Fig. 3). The same system has been recently explored by means of a stochastic, bit-strings model [28].

In this article, we explore how spatial correlations affect the dynamics described in Ref. [28], by extending the bit-strings model to a cellular automaton (CA). The CA model, albeit still being a minimal model including phenotypic heterogeneity, may be more adequate to investigate these dynamics in solid tumours. We are interested in the effects of space in the dynamics and transitions of the system. Moreover, our approach allows us to investigate the impact of cancer therapies targeting different phenotypes. In this sense, surgery, chemotherapy, and radiotherapy are the mainly used therapies against cancer. Their main limitation is the lack of selectivity for cancer cells over healthy cells, which results in insufficient drug or irradiation concentration in cancer cells, system toxicity, and the appearance of tumour cell populations resistant to drugs [29]. Targeted therapy aims to avoid toxicity for non-targeted cells and the evolution of drug-resistant tumours due to its specificity towards tumour cells. For example, this therapy can be used to deliver cytotoxic drugs to specific tumour cells carrying some aberrant receptors or presenting some anomalies at the surface of the cell. Its effectiveness is based on the targeted release at the site of the disease, reducing side effects affecting healthy tissues. Targeted cancer therapies mainly consist of monoclonal antibodies or small molecules with cytotoxic effects [29]. Monoclonal antibodies (i.e., antibodies that are made by identical immune cells) bind to specific antigens existing on the cell membrane, like transmembrane receptors, of tumour cells. Therapeutic monoclonal antibodies are associated to radio-isotopes or toxins that can deliver these cytotoxic agents into the targeted tumour cells [29]. For instance, a new anti-cancer therapy for colon cancer targets tumour cell-surface proteins with standard chemotherapy. It consists of a combination of monoclonal antibodies, prepared to bind to the tumour antigen LY6D/E46, and a common chemotherapy drug CPT-11, used for colorectal cancer [30].

Other approaches take advantage of the selective properties of viral infection by means of the so-called oncolytic viruses [31–34]. Also, a recent bio-inspired cancer drug delivery system, consisting of a nanocarrier made of pH-dependent self-degradable DNA (i.e., DNA nanoclew including DNAses on the surface), may be used to deliver small molecules inside tumour cells designed to alter their enzymatic activity [29]. For instance, release of the anti-cancer drug doxorubicin (DOX) has

a

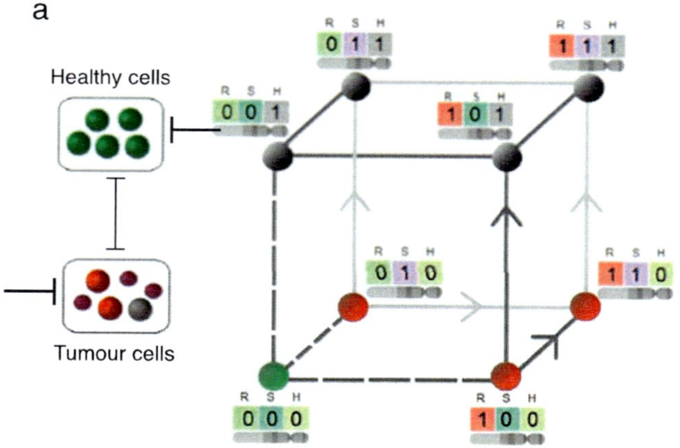

b

Cell state (R S H)	Variable	Phenotype: Replication	Phenotype: Mutation	Targeted therapy: Rate	Targeted therapy: Cell state
0 0 0	x_0	r	0	—	—
0 0 1	x_1	0	—	—	0 0 1
0 1 0	x_2	r	$\mu_b+\delta_\mu$	ε_2	0 1 1
0 1 1	x_3	0	—	—	0 1 1
1 0 0	x_4	$r+\delta_r$	μ_b	ε_4	1 0 1
1 0 1	x_5	0	—	—	1 0 1
1 1 0	x_6	$r+\delta_r$	$\mu_b+\delta_\mu$	ε_6	1 1 1
1 1 1	x_7	0	—	—	1 1 1

Fig. 1 Schematic diagram of the dynamical system investigated in this article. (**a**) Healthy cells (green marbles) compete with a heterogeneous pool of cancer cells (red marbles). The phenotype of each cell is coded by a 3-bit string. Each bit corresponds to a different compartment of genes: replication-related genes (1st bit, compartment R), genomic stability compartment (2nd bit, compartment S), and the house-keeping genes (3rd bit, compartment H). Healthy cells are coded with sequence 000 since they have no mutations or anomalies in these compartments. Note the full system lives in a three-dimensional sequence space. (**b**) Table with the cell state, the variable used for the differential equations model (the sub-index is the integer transformation of the binary number), and the phenotype (given by proliferation and mutation rates). The last two columns display the transition rates of the targeted therapy and the changes towards lethal sequences due to therapy

been recently studied in *in vitro* experiments using an acid environment (such as the one of internalised lysosomes), where the nanocapsules self-degrade and release the DOX [35].

2 Mathematical and Computational Models of Cancer Phenotypic Quasispecies

In this section, we first introduce two different models that have been used to investigate the system shown in Fig. 1: a deterministic model given by ordinary differential equations (see [26, 27] for details) and a stochastic agent-based model based on bit strings [28]. Then, we will introduce the new model given by a spatial version of the agent-based model. The dynamical system consists of a population of healthy cells that competes with a pool of cancer cells. Each cell state is coded by a binary string of length $\nu = 3$. Bit-strings simulation models are useful tools to investigate in silico evolution and have been employed to characterise the evolutionary dynamics of RNA viruses [24, 28, 36, 37], cancer quasispecies [18, 20], and genetic algorithms [38]. In our approach, each bit corresponds to a different compartment containing replication-related genes (compartment R, 1st bit), genes responsible for genomic stability (compartment S, 2nd bit), and a compartment with house-keeping genes (compartment H, 3rd bit), see Figs. 1 and 3.

For example, cell populations with increased proliferation rates (mutation on 1st bit or compartment R) are the result of anomalies in tumour-suppressor genes, such as TP53 (tumour protein 53) or MYC (L-myc, N-myc, and c-myc genes), and proto-oncogenes, for instance, RAS (Rat sarcoma genes family) or SRC (tyrosine-protein kinase Src) [3, 4, 39]. Both tumour-suppressor genes and proto-oncogenes induce tumourigenesis by increasing tumour cells population through, e.g., the stimulation of cell replication or the resistance to programmed cell death (apoptosis). Concerning genomic stability, several genes control the DNA repairing processes that, in case of mistakes produced during DNA replication or small DNA breaks, can act at different stages to arrange the DNA. Some of these genes are BRCA1 (breast cancer 1 gene), BLM (Bloom's syndrome gene), and ATM (ataxia telangiectasia-mutated), among others [4, 39]. Finally, the house-keeping genes (3rd bit or compartment H) are required for the maintenance of basic cellular function and cell survival. The house-keeping genes are expressed in all cells of an organism, i.e., ubiquitin, GAPDH, or ribosomal proteins, and an anomaly in any of these genes would result in cell death [4, 39].

Compartments with mutations or genetic anomalies are coded by bit 1, while bit 0 codes no anomalies. This system lives in a sequence space of dimension $2^{\nu=3}$. Hence, different phenotypes are found: the healthy phenotype (with sequence 000) and the pool of cancer cells with proliferative phenotypes, phenotypes with increased mutation rates (increased genomic instability), phenotypes with large proliferation and mutation rates, and the lethal phenotypes (see table (b) in Fig. 1).

2.1 Differential Equations Model

The mathematical model describing the population dynamics of healthy cells competing with the heterogeneous pool of cancer cells' phenotypes is based on Eigen's quasispecies equation [21] and considers a well-mixed and constant population of cells [26, 27] (see Fig. 1). The dynamical equations are

$$
\begin{aligned}
\dot{x}_{000} &= \frac{dx_0}{dt} = rx_0 - x_0\,\Phi(\mathbf{x}),\\
\dot{x}_{001} &= \frac{dx_1}{dt} = -x_1\,\Phi(\mathbf{x}),\\
\dot{x}_{010} &= \frac{dx_2}{dt} = (1-(\mu+\delta_\mu))rx_2 - x_2\,\Phi(\mathbf{x}),\\
\dot{x}_{011} &= \frac{dx_3}{dt} = \frac{1}{\nu-1}(\mu+\delta_\mu)rx_2 - x_3\,\Phi(\mathbf{x}),\\
\dot{x}_{100} &= \frac{dx_4}{dt} = (1-\mu)(r+\delta_r)x_4 - x_4\,\Phi(\mathbf{x}),\\
\dot{x}_{101} &= \frac{dx_5}{dt} = \frac{1}{\nu-1}\mu(r+\delta_r)x_4 - x_5\,\Phi(\mathbf{x}),\\
\dot{x}_{110} &= \frac{dx_6}{dt} = (1-(\mu+\delta_\mu))(r+\delta_r)x_6\\
&\quad + \frac{1}{\nu-1}\Big[(\mu+\delta_\mu)rx_2 + \mu(r+\delta_r)x_4\Big] - x_6\,\Phi(\mathbf{x}),\\
\dot{x}_{111} &= \frac{dx_7}{dt} = (\mu+\delta_\mu)(r+\delta_r)x_6 - x_7\,\Phi(\mathbf{x}),
\end{aligned}
\tag{1}
$$

where $\Phi(\mathbf{x}) = r(x_0+x_2)+(r+\delta_r)(x_4+x_6)$ is the dilution flow, which introduces competition between all the cell populations, also keeping population constant. The state variables $x_i(t)$, $i = 0,\ldots,7$, are the relative concentration or population numbers of cells with sequences 000, ..., 111. Here, sub-indices of the population variables correspond to the integer number of the binary sequences at time t. The model parameters are given by the replication rate of cells ($r > 0$); the increase of proliferation of tumour cells ($\delta_r > 0$); the rate of mutation or accumulation of genomic anomalies of tumour cells $0 < \mu < 1$; and the increase of genome instability of tumour cells $0 < \delta_\mu < 1-\mu$. That is, tumour cells with bit 1 in the first position of the string present increased proliferation rates, $r+\delta_r$, due to mutations or anomalies in replication-related genes. Genome instability is introduced with $\mu+\delta_\mu$ when the second bit of the strings is 1. The cells with sequences $ab1$, $a, b \in \{0, 1\}$ present anomalies or mutations in hk genes and thus are not able to proliferate (see Fig. 1). Finally, ν is the length of the sequences (here with $\nu = 3$).

The fixed points of Eqs. (1) and their stability were characterised in [26] (see also [27]). Among the fixed points identified, two of them were responsible for the two asymptotic states behind tumour persistence and extinction. A summary of the dynamics is shown in Fig. 2. Panel (a) shows how the equilibrium of the population of cells changes at increasing tumour cells mutation rate. Here, two well-defined phases are found: (i) survival of tumour cells (cells 100, 110, 101, and 111) and extinction of healthy cells; (ii) outcompetition of tumour cells by the healthy populations (healthy state). That is, there exists an upper limit of the mutation rate beyond which tumour cells become extinct, in agreement with previous cancer quasispecies models [18–20]. As a difference from these previous works, the bifurcation responsible for this shift is discontinuous (see below). These two different regimes correspond to different stability scenarios for two fixed points, labelled P_2^* and P_3^* in [26] and given by

$$P_2^* = (1, 0, 0, 0, 0, 0, 0, 0),$$

$$P_3^* = \left(0, 0, 0, 0, \frac{2(1-\mu)\delta_\mu}{\mu + 2\delta_\mu}, \frac{\mu\delta_\mu}{\mu + 2\delta_\mu}, \frac{\mu(1-\mu)}{\mu + 2\delta_\mu}, \frac{\mu(\mu + \delta_\mu)}{\mu + 2\delta_\mu}\right).$$

According to the calculations developed in [26], when the mutation rate μ is below the bifurcation value $\mu_c = \delta_r/(r + \delta_r)$, the fixed point P_3^* is globally asymptotically stable, while P_2^* is unstable, meaning that tumour cells outcompete the healthy ones. This result can be observed in Fig. 2a by means of a bifurcation diagram. Figure 2b displays different time series below the bifurcation value (upper and middle panels) and one example of the extinction of the tumour cells once $\mu > \mu_c$ (lower panel). We note that P_2^* and P_3^* have a heteroclinic connection (see below for the definition of heteroclinic connection). Figure 2c shows three phase portraits with orbits projected on the phase space (x_0, x_4). At the bifurcation value $\mu = \mu_c$ (mid phase portrait), the heteroclinic connection is replaced by a line of equilibria (giving place to the so-called normally hyperbolic invariant manifold [27]).

After the bifurcation, when $\mu > \mu_c$, the stability of these two fixed points is reversed, P_2^* being asymptotically globally stable and P_3^* being unstable. This bifurcation, which is global, has been named as trans-heteroclinic bifurcation [27]. Such a bifurcation has also been identified in a quasispecies model describing the so-called survival-of-the-flattest effect [27, 40]. Hence, the dynamics of the system under study under its mean field limit involves monostability when $\mu \neq \mu_c$ [26]. This means that the stable fixed points are globally stable and neither coexistence of solutions involving stable populations of healthy and tumour cells nor different basins of attraction can be found in the phase space of Eq. (1) (see [26]).

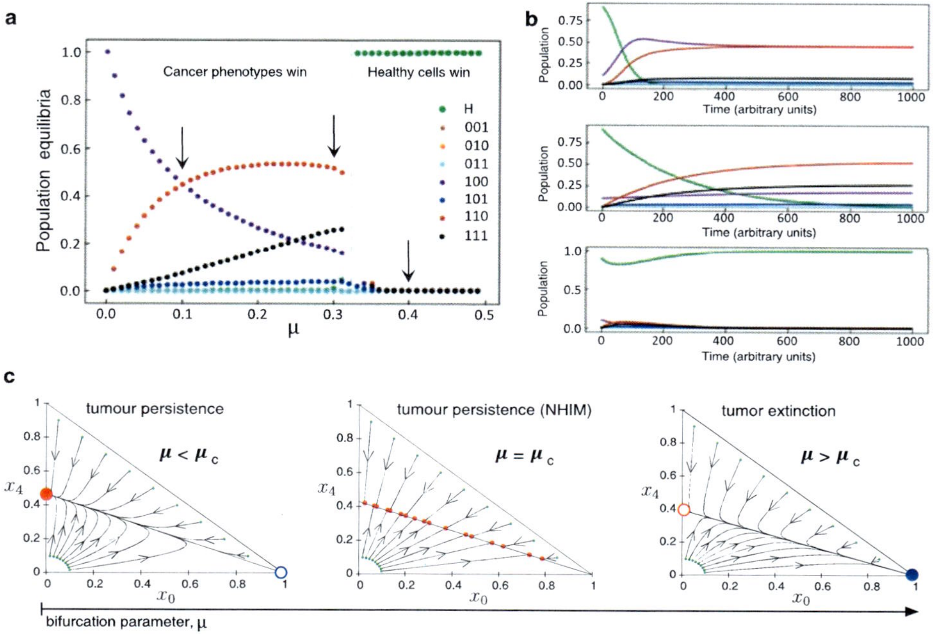

Fig. 2 Dynamics and bifurcations for Eq. (1). (**a**) Mean population equilibria at increasing mutation rate using as initial populations ($\tau = 0$) a 10% of tumour cells (with the proliferative phenotype C_{100}) and a 90% of healthy cells. Notice that at the value $\mu = \mu_c = 1/3$, the trans-heteroclinic bifurcation involves a catastrophic extinction of the tumour cells. (**b**) Time series of the population of cells computed with $\mu = 0.1$, $\mu = 0.3$, and $\mu = 0.4$. (**c**) Phase portraits show the orbits projected on the phase space (x_0, x_4). The left plot displays a case for $\mu < \mu_c$, for which the equilibrium point P_2^* is unstable (open blue dot). At $\mu = \mu_c$ (mid panel), the heteroclinic connection between the equilibria P_2^* and P_3^* is replaced by an attracting line (NHIM: Normally Hyperbolic Invariant Manifold), and the equilibrium values depend upon the initial conditions. Finally, the panel on the right displays the orbits for $\mu > \mu_c$, where equilibrium P_2^* is asymptotically, globally stable (solid blue dot) and P_3^* is unstable (open red dot). In all panels, we have used $r = 0.1$, $\delta_r = \delta_\mu = 0.05$

As mentioned, these two equilibrium points are heteroclinically connected (see also [27]). The so-called heteroclinic connection can be defined as follows:

Definition 1 (Heteroclinic Connection) Let x_1^* and x_2^* be equilibria of a nonlinear function $f : \mathbb{R} \to \mathbb{R}^n$

$$\dot{x}(t) = f(x(t)), \tag{2}$$
$$\text{s.t.} \quad x(0) = x_0 \in \mathbb{R}^n,$$

where f is continuously differentiable. An orbit $\text{Or}(x_0)$ starting at a point $x_0 \in \mathbb{R}^n$ is called a heteroclinic connection of x_1^* and x_2^* if $\lim_{t\to-\infty} \phi^t(x_0) = x_1^*$ and $\lim_{t\to\infty} \phi^t(x_0) = x_2^*$, $\phi^t(x_0)$ being the value of a trajectory starting at the point x_0 at time t.

Remark 1 From Definition 1, it immediately follows that the heteroclinic connection is a part of the unstable manifold of x_1^*, $W^u(x_1^*)$, as well as of the stable manifold of x_2^*, $W^s(x_2^*)$.

2.2 *Stochastic Bit-Strings Model*

To explore the role of finite population size, we introduce an agent-based probabilistic model in which cell phenotypic traits are also coded by bit strings (see Fig. 3). Specifically, we will describe the spatial model given by a Cellular Automaton (CA), in which each cell of the lattice is either a healthy or a tumour cell. The CA model can be easily transformed to the well-mixed stochastic system by breaking the spatial correlations, thus having exactly the same model of Ref. [28].

The CA model consists of a two-dimensional state space $\Theta(L^2)$, with a finite population of L^2 cells, being L the number of cells in one side of the lattice. Each cell is labelled as $C_{abc}^{(i,j)}$, with $i, j = 1, \ldots, L$, being spatial coordinates and $a, b, c \in \{0, 1\}$ coding the phenotype of each cell. In what follows, we will also alternatively use as sub-indices both the sequence and its integer value. Recall that a and b denote the replication-related and genomic stability compartments, respectively, while c denotes the house-keeping genes compartment (see Figs. 1 and 3). The simulation algorithm works as follows (see Fig. 3): at each time generation, τ, we randomly choose L^2 cells, say $C_{abc}^{(i,j)}$, of the population and then apply the state-transition rules described below. Each time a cell is chosen, we also choose a random neighbour using a Moore neighbourhood with $C_{abc}^{(i\pm k, j\pm l)}$, where $k, l \in \{0, 1\} \setminus \{k = l = 0\}$, i.e., 8 nearest neighbours. This asynchronous updating ensures that, on average, all

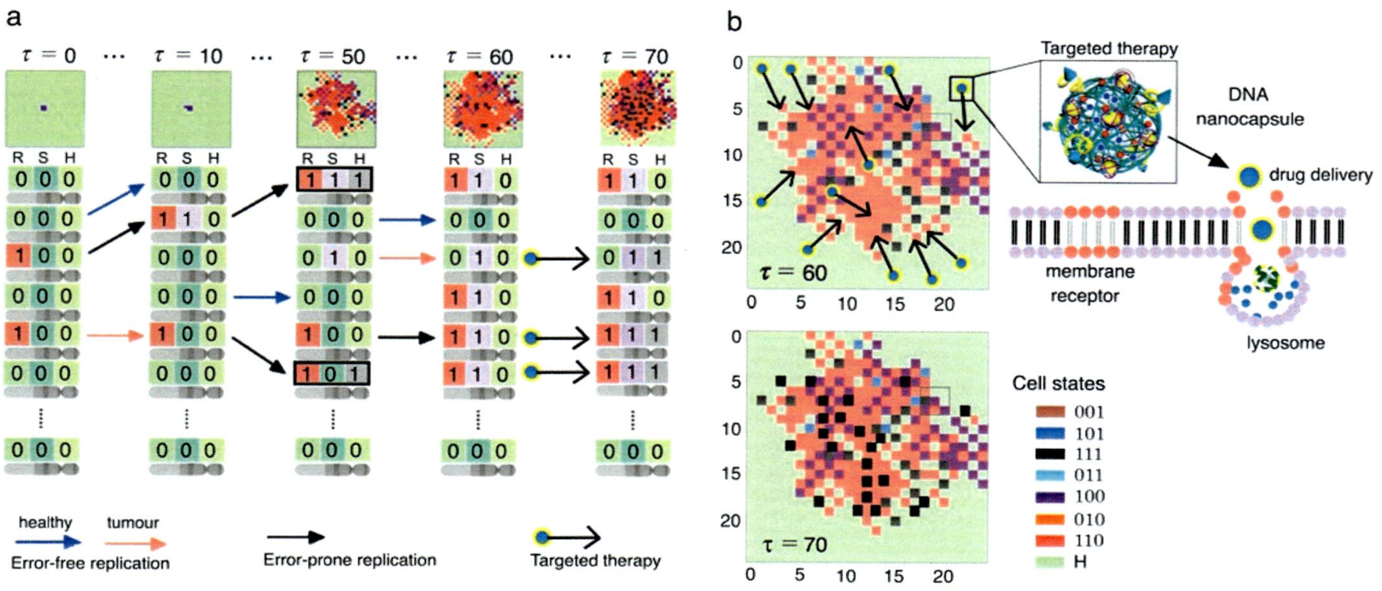

Fig. 3 (**a**) Schematic diagram of the stochastic cellular automaton (CA) model using bit strings. From a small initial population of tumour cells (time generation $\tau = 0$), the state-transition rules that consider proliferation events together with mutation are applied to the full population and the evolutionary dynamics is tracked. The last two columns display the application of cancer targeted therapy after generation $\tau = 60$, for which specific phenotypes are targeted (see panel (**b**) for a more detailed diagram on targeted therapy)

cells of $\Theta(L^2)$ will be updated once per generation. The state-transition rules are given by the following:

1. **Proliferation without mutation**. Healthy cells proliferate with probability $r \in [0, 1]$. Following [26], we assume that the mutation rate of healthy cells is negligible compared to the mutation rate of tumour cells. Hence, healthy cells reproduce according to the next reaction (blue arrows in Fig. 3a)

$$C_{000}^{(i,j)} + C_{abc}^{(i\pm k,j\pm l)} \xrightarrow{r} C_{000}^{(i,j)} + C_{000}^{(i\pm k,j\pm l)}.$$

Cancer cells can proliferate with error-free replication (solid red arrows in Fig. 3a) or making anomalous copies of themselves (black arrows in Fig. 3a). The error-free reaction of the cells with sequence C_{100} is

$$C_{100}^{(i,j)} + C_{abc}^{(i\pm k,j\pm l)} \xrightarrow{(1-\varepsilon_4)\,(r+\delta_r)\,(1-\mu_b)^{\nu}} C_{100}^{(i,j)} + C_{100}^{(i\pm k,j\pm l)}.$$

Since cells with state 100 have anomalies in the replication-related genes, they reproduce with probability $(r+\delta_r) \in [0, 1]$, δ_r being the increase in proliferation rates tied to the anomalies found in the replication compartment. The terms $\varepsilon_{4,2,6}$ will only apply when considering targeted therapies, being 0 without therapy (see below). Here, $\mu_b \in [0, 1]$ is the per-bit mutation probability of tumour cells and ν is the length of the string (here $\nu = 3$). Probability μ_b can also be considered as a probabilistic rate at which cells accumulate damage at each compartment (by means of, e.g., gene loss, chromosomal breaks). For the other two proliferative phenotypes, one has

$$C_{010}^{(i,j)} + C_{abc}^{(i\pm k,j\pm l)} \xrightarrow{(1-\varepsilon_2)\,r\,(1-(\mu_b+\delta_\mu))^{\nu}} C_{010}^{(i,j)} + C_{010}^{(i\pm k,j\pm l)},$$

$$C_{110}^{(i,j)} + C_{abc}^{(i\pm k,j\pm l)} \xrightarrow{(1-\varepsilon_6)\,(r+\delta_r)\,(1-(\mu_b+\delta_\mu))^{\nu}} C_{110}^{(i,j)} + C_{110}^{(i\pm k,j\pm l)}.$$

2. **Proliferation with mutation**. The reactions for erroneous replication apply to the proliferative tumour cells, according to

$$C_{100}^{(i,j)} + C_{abc}^{(i\pm k,j\pm l)} \xrightarrow{(1-\varepsilon_4)\,(r+\delta_r)\,(1-\mu_b)^{\nu-1}\,(\mu_b)^{\nu-2}} C_{100}^{(i,j)} + C_{100}^{(i\pm k,j\pm l)},$$

here with $b = 1$ and $c = 0$, or $b = 0$ and $c = 1$ (recall that here no backward mutations are allowed, following [26]). For cells with sequence 010, one has

$$C_{010}^{(i,j)} + C_{abc}^{(i\pm k,j\pm l)} \xrightarrow{(1-\varepsilon_2)\,r\,(1-(\mu_b+\delta_\mu))^{\nu-1}(\mu_b+\delta_\mu)^{\nu-2}} C_{010}^{(i,j)} + C_{p1m}^{(i\pm k,j\pm l)},$$

with $p = 1$ and $m = 0$ or $p = 0$ and $m = 1$. Here, δ_μ is the increase in mutation probabilities due to the anomalies tied to the compartment of genome instability.

The other reproducing tumour cells have state 110, with reaction

$$C_{110}^{(i,j)} + C_{abc}^{(i\pm k,j\pm l)} \xrightarrow{(1-\varepsilon_6)\,(r+\delta_r)\,(1-(\mu_b+\delta_\mu))^{\nu-1}\,(\mu_b+\delta_\mu)^{\nu-2}} C_{110}^{(i,j)} + C_{111}^{(i\pm k,j\pm l)}.$$

Finally, the cells with anomalies in the hk genes compartment do not proliferate, i.e., S_{ab1}, with $a, b \in \{0, 1\}$.

3. **Targeted therapies against phenotypes**. Concerning the implementation of cancer targeted therapies, we will introduce the following reactions to evaluate how specificity in lethality influences dynamics. We will assume that therapy is selective and acts upon the following proliferative tumour phenotypes:

$$C_{010}^{(i,j)} \xrightarrow{\varepsilon_2} C_{011}^{(i,j)}, \qquad C_{100}^{(i,j)} \xrightarrow{\varepsilon_4} C_{101}^{(i,j)}, \qquad C_{110}^{(i,j)} \xrightarrow{\varepsilon_6} C_{111}^{(i,j)}. \tag{3}$$

Here, $\varepsilon_{2,4,6}$ corresponds to the integer number from the binary sequences 010, 100, and 110, respectively.

Let us define the fraction of healthy cells (normalised population) as

$$H = \frac{1}{L^2}\sum_{i=1}^{L}\sum_{j=1}^{L} C_{000}^{(i,j)}.$$

The normalisation of the populations allows us to define a state space for this stochastic dynamical system given by a seven-dimensional simplex

$$\Omega := \left\{ \tilde{C}_\kappa = 0, \ldots, 2^{\nu-1} \in \mathbb{R}^{2^\nu-1} \;\middle|\; \tilde{C}_\kappa = \frac{1}{L^2}\sum_{i=1}^{L}\sum_{j=1}^{L} C_\kappa^{(i,j)} \geq 0, \sum_{\kappa=0}^{2^\nu-1} \tilde{C}_\kappa = 1 \right\},$$

κ being the integer number of the binary sequence. We note that we will use alternatively the two notations with the full sequence or the integer transformation of the binary sequence. The dynamics in terms of population asymptotic states as well as in transients will be mainly analysed using projections of Ω for several initial conditions or for a single initial population. If not otherwise specified, we will consider an initial population with 90% of healthy cells and 10% of tumour cells with driver mutations (anomalies in compartment R), i.e., $\tilde{C}_{000}^{(i,j)}(\tau = 0) = 0.9$, $\tilde{C}_{100}^{(i,j)}(\tau = 0) = 0.1$, and all other sequences to zero, as initial conditions. These initial conditions assume that a small number of tumour cells initiate the process of tumourigenesis, as it may happen in real tumours.

For simplicity, and to better analyse the impact of spatial correlations, we will use large lattice sizes to minimise the impact of noise, which was studied in detail in Ref. [28] for the well-mixed system. The consideration of large stochastic fluctuations involved a change from the discontinuous transition given by the trans-heteroclinic bifurcation to a smooth one, allowing for the emergence of noise-induced bistability [28]. We will hereafter set $L = 250$ (if not otherwise specified), thus having a total population of 62,500 cells.[1] For the sake of clarity and comparison, the results presented below will display both analyses for the well-mixed stochastic system and the CA model, first without considering targeted therapy (i.e., $\varepsilon_{2,4,6} = 0$). Then, in the next section, we will introduce the results by considering different targeted therapies. We recall that the well-mixed dynamics have been simulated by breaking all neighbour-dependent processes. That is, instead of choosing a random neighbour within the 8 nearest cells, the model randomly chooses any cell of the lattice as a neighbour.

The spatial simulations reveal that the bifurcation value, compared to the well-mixed stochastic system (and under the selected probability values and large population sizes), remains similar (see Fig. 4). However, the stationary distributions of the tumour phenotypes change. For the well-mixed system, the tumour cells that persist below the critical per-bit mutation probability, μ_b^c, are the ones given by the equilibrium point P_3^* (compare the diagrams of Figs. 2a and 4a). The same analysis for the CA model shows a lower extinction threshold due to mutation for sequences 100 (violet dots) and 101 (blue dots). This phenomenon involves a much higher stationary values of the proliferative and mutagenic phenotypes (sequence 110) as well as its lethal variant 111. This means that the fastest proliferative cells (100) actually accumulate much more mutations (due to increasing μ_b), giving place to a dominant proliferative cell with increased genomic instability in solid tumours, since this effect is not observed in the non-spatial simulations. The transition of these phenotypes still remains discontinuous, and once $\mu_b > \mu_b^c$, the healthy cells outcompete the tumour cells.

Another interesting, yet expected result, is that transient times are much longer in the spatial simulations. Figure 5a, b displays time series for three different values of per-bit mutation probabilities (setting all other probabilities at $r = 0.1$ and $\delta_{r,\mu} = 0.05$). Two of them below the critical per-bit mutation with $\mu = 0.05$ (upper panels) and $\mu = 0.15$ (middle panels), plus another one with a mutation above the critical value, i.e., $\mu = 0.2$ (lower panels). Notice that in all panels the time needed to achieve stationarity is much longer for the spatial simulations. Figure 5b shows different snapshots with the spatial distributions of the cells at different generation times for the panels with $\mu = 0.15$ (b1–b4) and $\mu = 0.2$ (b5–b8). Note that for the value above the bifurcation, the tumour cell populations vanish after an initial growth, being outcompeted by the healthy cells (represented in green colour).

Finally, Fig. 6 displays stochastic trajectories projected in the same phase space displayed in Fig. 2c, using $(H, \tilde{C}_{100})$, and recall that $\tilde{C}_{100} = L^{-2} \sum_{i=0}^{L} \sum_{j=0}^{L} C_{100}^{(i,j)}$ is the (normalised) population of cells with sequence 100 in the entire lattice. We

[1]This amount of cells reproduced the results of the ODEs model [27]. Nevertheless, smaller population sizes will also be used to show dynamics and the emergence of noise-induced bistability [28], also found in the CA model.

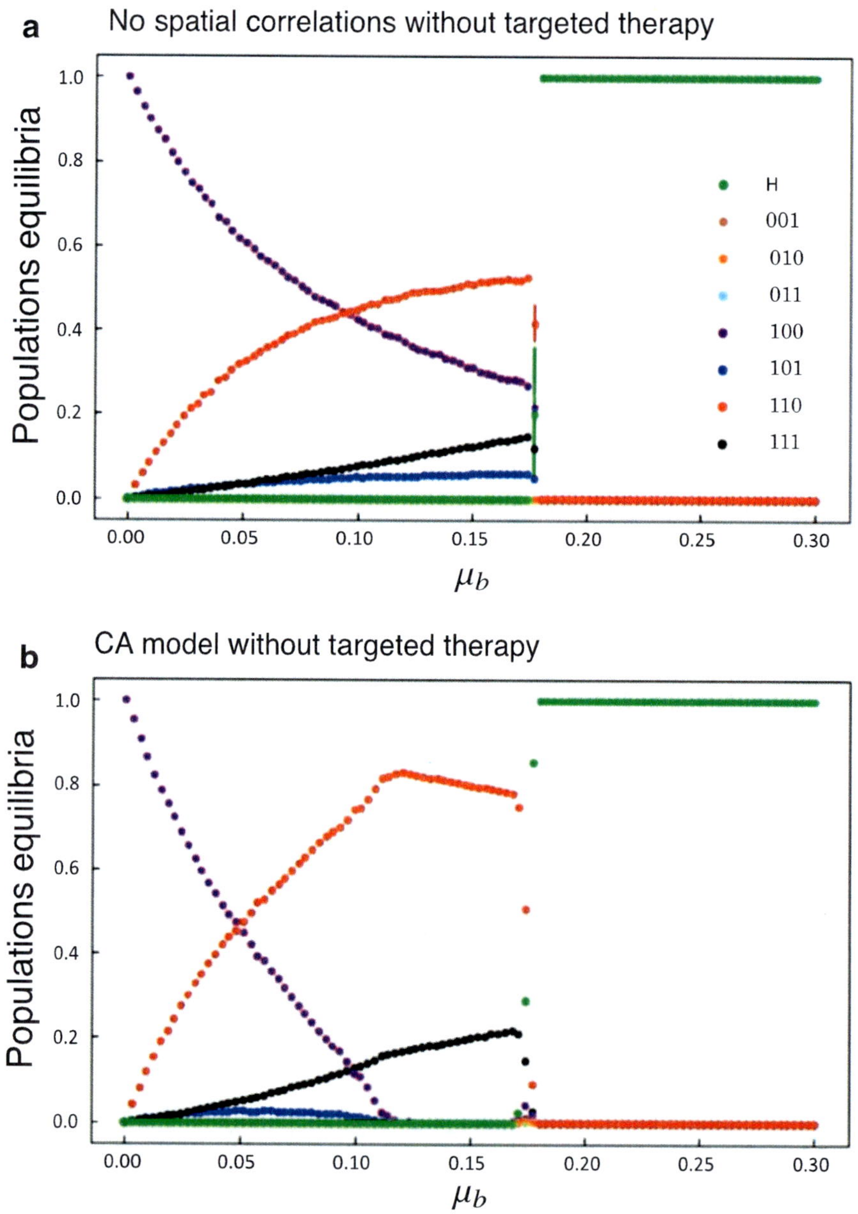

Fig. 4 Equilibrium values of cells at increasing per-bit mutation probability (μ_b) for the bit-string model without (**a**) and with (**b**) spatial correlations using $r = 0.1$, and $\delta_{r,\mu} = 0.05$. We plot the population equilibria (mean ± SD) of the different phenotypes considering $N = 62{,}500$ cells ($L = 250$) obtained at $\tau = 25{,}000$ generations and averaged over 5 independent replicas. The initial conditions consist of an initial population of 400 cells with sequence C_{100} and all other cells being healthy. For the CA model, the tumour cells are placed in the middle of the lattice in a 20×20 square

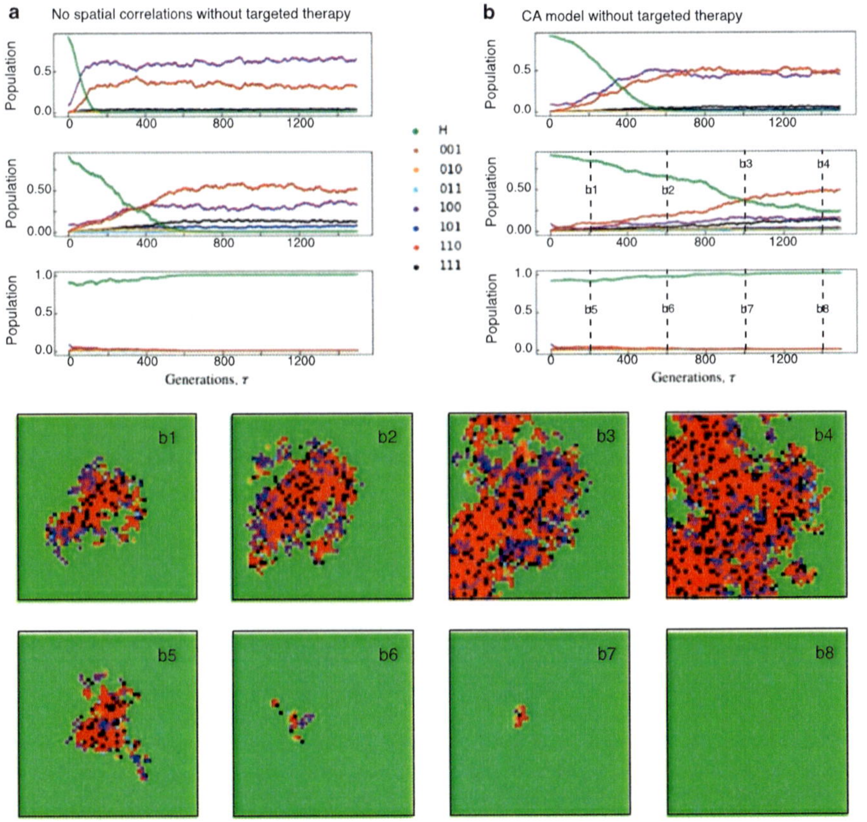

Fig. 5 (**a**, **b**) Time series obtained for the stochastic bit-string models without (**a**) and with (**b**) spatial correlations with $\mu_b = 0.05$ (upper), $\mu_b = 0.15$ (middle), and $\mu_b = 0.2$ (lower), where $\mu_b^c \approx 0.18$. Note that transients are much longer for the CA simulations. For the spatial system, we display snapshots corresponding to the simulations before (mid panel in (**b**)) and after (lower panel in (**b**)) the critical mutation probability. Each snapshot is identified in the corresponding time series above. The different cell populations are displayed with the same colours than the ones used in the time series. Here, we have used a smaller population of 50×50 cells and a 10% of tumour cells (250 cells) with sequence C_{100} as an initial condition. For the CA model, the tumour populations have been placed at the centre of the lattice Θ. The other probabilities are fixed as in Fig. 4

note that the heteroclinic connection still remains visible in both well-mixed and spatial simulations. Here, we have used a smaller population size ($L^2 = 2500$ cells), and the third panel of the first row displays the so-called noise-induced bistability. This phenomenon involves that either the healthy or the tumour states can be achieved due to stochastic fluctuations (see [28] for details of the well-mixed system).

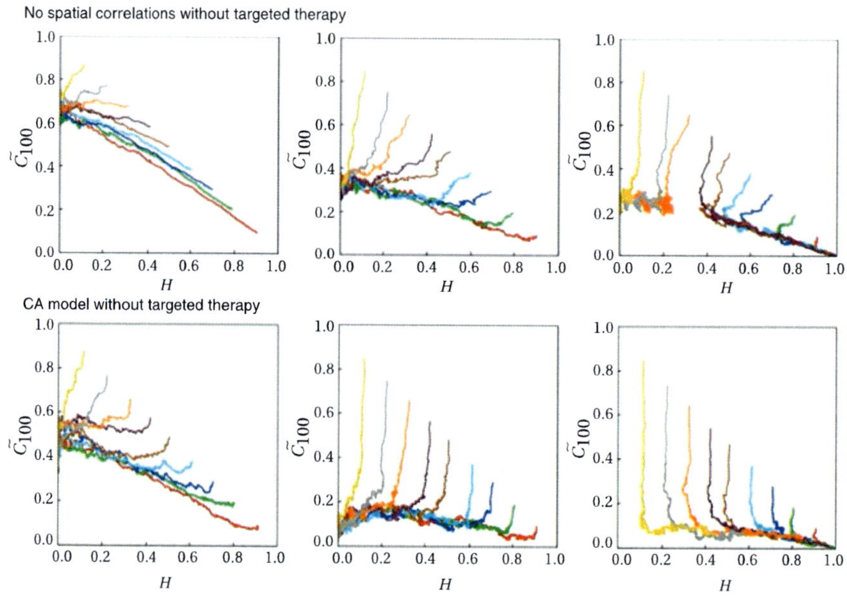

Fig. 6 Phase portraits of the stochastic models without (upper row) and with (lower row) spatial correlations. The stochastic trajectories (displayed with different colours) are projected on the space $\Omega := (H, \tilde{C}_{100})$, being $\tilde{C}_{100} = L^{-2} \sum_{i=0}^{L} \sum_{j=0}^{L} C_{100}^{(i,j)}$. Panels at the left are built setting $\mu_b = 0.05$, while panels at the middle and at the right use $\mu_b = 0.15$ and $\mu_b = 0.2$, respectively (all other parameters are kept as in the previous figure). Here, we have used $L = 50$ to illustrate the phenomenon of noise-induced bistability (see right panel for the non-spatial simulations). Notice that the heteroclinic connection remains visible in all phase portraits

2.3 *The Impact of Targeted Therapy*

One of the main interests of our model is to analyse the impact of selective therapies against cancer phenotypes. As mentioned, several therapeutic strategies based on the selective targeting of tumour cells are being currently studied. For instance, oncolytic viruses [31–34] or nanoparticles carrying different cytotoxic agents such as doxorubicine [35]. Despite the target of cancer cells is not a trivial process, especially at the phenotypic level, we here investigate the impact of both single and combined targeted therapies against different cancer phenotypes (see Fig. 3b). To do so, we will assume that tumour cells with a given phenotype are targeted by, e.g., nanoparticles, causing cytotoxic or lethal mutagenic effects, thus changing the sequence to a lethal phenotype (this is implemented by changing the third bit according to the reactions (3) shown in Sect. 2.2).

Figure 7 displays how the equilibrium populations of the different cells' phenotypes change at increasing μ_b under different targeted therapies. We have first analysed the results of targeting single phenotypes, focusing on sequences 100, 010, and 110. Typically, the most effective strategy is to target the faster replicating

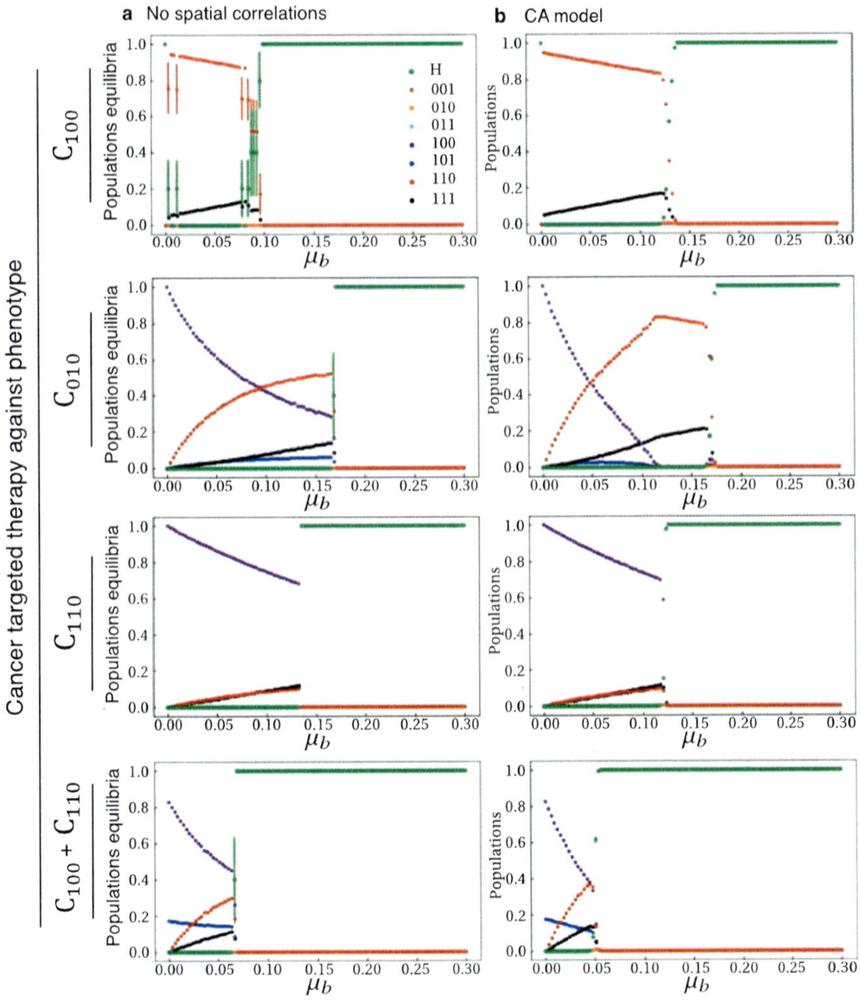

Fig. 7 Impact of therapeutic targeting against different phenotypes on the equilibrium populations of healthy and tumour cells. Each panel displays the mean (±SD) populations at increasing μ_b (computed as in Fig. 4). The first and second columns show results for the well-mixed and spatial systems, respectively. The first three rows display the results for single target, while the fourth one displays a combination therapy targeting phenotypes 100 and 110 simultaneously. For the single targeting, we have set $\varepsilon_{2,4,6} = 0.1$, while for the combined one we use $\varepsilon_{4,6} = 0.025$. In all panels, we have used $L^2 = 62{,}500$ cells and all other probability values are kept as in the previous figures

phenotypes, for which the critical per-mutation probability, μ_b^c, largely diminished, thus indicating that the tumour cells become much more fragile to increases in mutation (see first and third rows in Fig. 7). Interestingly, the targeting of sequence 010 does not show a clear change in μ_b^c. On the contrary, the targeting of sequences 110 displays a lower μ_b^c value.

As previously discussed, the difference between the non-spatial and spatial simulations is that for some targeting strategies the stationary distribution of the cells can change. We note that the rates of lethality associated to single targeted therapy have been set at $\varepsilon_{2,4,6} = 0.1$. A combined therapy against sequences containing the most replicative phenotypes, i.e., 100 + 110, shows a huge decrease of μ_b^c, which is placed at $\mu_b^c \approx 0.07$ for the non-spatial model and at $\mu_b^c \approx 0.05$ for the spatial one. These results have been obtained by setting $\varepsilon_{4,6} = 0.025$, which, compared to the single targeting, is a much lower probability of shifting replicating tumour cells to lethal ones. Indeed, the resulting scenario with lethality rates due to the therapy ($\varepsilon_4 = \varepsilon_6 = 0.1$) is the dominance of the healthy cells for any value of the per-bit mutation rate (results not shown). This suggests a strong synergy arising from combined therapy, which seems to be amplified in the spatial system.

3 Conclusions

We have investigated the dynamics of cancer phenotypic quasispecies considering space in an explicit manner. This system, studied in Refs. [26, 28] using well-mixed approaches, considers a heterogeneous population of tumour cells competing with healthy ones. The deterministic mean field model has been shown to exhibit a population shift ruled by a trans-heteroclinic bifurcation [27] when the mutation rate of cancer cells overcomes a critical mutagenic threshold. The asymptotic dynamical outcomes when the mutation rate is below this critical mutagenic limit are the dominance of tumour cells and the outcompetition of healthy cells. Once the mutagenic threshold is achieved, the situation is reversed. Healthy cells outcompete the pool of tumour phenotypes, and the systems enter into a "healthy" state.

Concerning the stochastic bit-strings model without spatial correlations, the bifurcation diagram also displays the trans-heteroclinic bifurcation identified for the mean field model [28]. The main difference between the mean field model and the stochastic system simulated using bit strings is the change from a discontinuous extinction of the tumour cells to a continuous one [28]. Also, the stochastic model displays the so-called noise-induced bistability, not found in the differential equations model since no bistability was found.

When including spatial correlations to the stochastic bit-strings system with the CA model, the behaviour of the quasispecies population dynamics is altered. The mean population equilibria values are modified, and typically the transient times become longer. For large population sizes (large L), the trans-heteroclinic bifurcation is only observed for the tumour phenotypes with sequences 110 and 111, while tumour cells with sequence 100 exhibit a progressive population descent occurring before the trans-heteroclinic bifurcation takes place at increasing μ_b.

The impact of targeted cancer therapies on tumour extinction thresholds has also been analysed. On one hand, the single targeting of phenotypes with mutations in the replication-related genes (1st bit) or with anomalies in both replication-related genes and genes responsible for genomic stability (1st and 2nd bits) decreases the value of the critical per-bit mutation significantly. On the other hand, the target of phenotypes with only anomalies in the genetic stability compartment did not produce significant decrease of μ_b. However, when applying a combined therapy targeting simultaneously both highly proliferative phenotypes (100 + 110), the effectivity of the therapy is drastically magnified.

The approaches investigated in this article as well as in Refs. [26, 28] aim at introducing minimal models considering genotypic–phenotypic variability in the population of tumour cells. This is a very simplistic picture of cancer dynamics because we are not considering important features like tumour microenvironment, supply of nutrients and oxygen by angiogenesis, or the effect of the immune system. As mentioned in the Introduction, sources of phenotypic variability might not be a direct consequence of changes at the genotypic or karyotypic level. Our model could be extended including these features together with the intrinsic variability due to mutations or genomic anomalies.

Acknowledgments This work has been partially funded by a MINECO grant MTM-2015-71509-C2-1-R and the Spain's "Agencia Estatal de Investigación" grant RTI2018-098322-B-I00. JS has also been funded by a "Ramón y Cajal" Fellowship (RYC-2017-22243).

References

1. Gerlinger, M., Rowan, A.J., Horswell, S., Larkin, J., et al.: Intratumor heterogeneity and branched evolution revealed by multiregion sequencing. N. Engl. J. Med. **366**, 883–892 (2012)
2. Cusnir, M., Cavalcante, L.: Inter-tumor heterogeneity. Hum. Vaccin. Immunother. **8**(8), 1143–1145 (2012)
3. Hanahan, D., Weinberg, R.A.: The hallmarks of cancer. Cell **100**, 57–70 (2000)
4. Hanahan, D., Weinberg, R.A.: Hallmarks of cancer: the next generation. Cell **144**, 646–670 (2011)
5. Hinohara, K., Polyak, K.: Intratumoral heterogeneity: more than just mutations. Trends Cell Biol. **29**(7), 569–570 (2019)
6. Anderson, G.R., Stoler, D.L., Brenner, B.M.: Cancer as an evolutionary consequence of a destabilised genome. Bioessays **23**, 103746 (2001)
7. Loeb, L.A.: A mutator phenotype in cancer. Cancer Res. **61**, 3230–3239 (2001)
8. Loeb, L.A.: Human cancers express the mutator phenotypes: origin, consequences and targeting. Nature **11**, 450–457 (2011)
9. Lengauer, C., Kinzler, K.W.: Vogelstein, B. Genetic instabilities in human cancers. Nature **396**, 643–649 (1998)
10. Cahill, D.P., Kinzler, K.W., Vogelstein, B., Lengauer, C.: Genetic instabilities and Darwinian selection in tumors. Trends Genet. **15**, M57–M61 (1999)
11. Merlo, L.M.F., Pepper, J.W., Reid, B.J., Maley, C.C.: Cancer as an evolutionary and ecological process. Nat. Rev. Cancer **6**, 924–935 (2006)
12. Sniegowski, P.D., Gerrish, P.J., Lenski, R.E.: Evolution of high mutation rates in experimental populations of *E. coli*. Nature **387**, 703–705 (1997)
13. Bielas, J.H., Loeb, K.R., Rubin, B.P., True, L.D. et al.: Human cancers express a mutator phenotype. Proc. Natl. Acad. Sci. U.S.A. **103**, 18238–8242 (2006)
14. Fox, E.J., Loeb, L.A.: Lethal mutagenesis: targeting the mutator phenotype in cancer. Semin. Cancer Biol. **20**, 353–359 (2010)

15. Loeb, L.A.: Human cancers express the mutator phenotypes: origin, consequences and targeting. Nature **11**, 450–457 (2011)
16. Oliver, A., Canton, R., Campo, P., Baquero, F. et al.: High frequency of hypermutable *Pseudomonas aeruginosa* in cystic fibrosis lung infection. Science **288**, 1251–1254 (2000)
17. Bjedov, I., Tenaillon, O., Gérard, B., Souza, V. et al.: Stress-induced mutagenesis in bacteria. Science **300**, 1404–1409 (2003)
18. Solé, R.V.: Phase transitions in unstable cancer cell populations. Eur. J. Phys. B **35**, 117–124 (2003)
19. Solé, R.V., Deisboeck, T.S.: An error catastrophe in cancer? J. Theor. Biol. **228**, 47–54 (2004)
20. Solé, R.V., Valverde, S., Rodríguez-Caso, C., Sardanyés, J.: Can a minimal replicating construct be identified as the embodiment of cancer? Bioessays **36**, 503–512 (2014)
21. Eigen, M.: Self-organization of matter and the evolution of biological macromolecules. Naturwiss **58**, 465–523 (1971)
22. Eigen, M., McCaskill, J., Schuster, P.: The molecular quasi-species. Adv. Chem. Phys. **75**, 149–263 (1989)
23. Bull, J.J., Meyers, L.A., Lachmann, M.: Quasispecies made simple. PLOS Comput. Biol. **1**(6), e61 (2005)
24. Solé, R.V., Sardanyés, J., Díez, J., Mas, A.: Information catastrophe in RNA viruses through replication thresholds. J. Theor. Biol. **240**(3), 353–359 (2006)
25. Sardanyés, J., Simó, C., Martínez, R., Solé, R.V., Elena, S.F.: Variability in mutational fitness effects prevents full lethal transitions in large quasispecies populations. Sci. Rep. **4**, 4625 (2014)
26. Sardanyés, J., Martínez, R., Simó, C., Solé, R.: Abrupt transitions to tumor extinction: a phenotypic quasispecies model. J. Math. Biol. **74**(7), 1589–1609 (2017)
27. Sardanyés, J., Simó, C., Martínez, R.: Trans-heteroclinic bifurcation: a novel type of catastrophic shift. R. Soc. Open Sci. **5**, 171304 (2018)
28. Sardanyés, J., Alarcón, T.: Noise-induced bistability in the fate of cancer phenotypic quasispecies: a bit-strings approach. Sci. Rep. **8**, 1027 (2018)
29. Padma, V.V.: An overview of targeted cancer therapy. BioMedicine **5**, 19 (2015)
30. Lane, D.: Designer combination therapy for cancer. Nat. Biotech. **24**, 163–164 (2006)
31. Post, D.E., Shim, H., Toussaint-Smith, E., Van Meir, E.G.: Cancer scene investigation: how a cold virus became a tumor killer. Future Oncol. **1**, 247–258 (2005)
32. Roberts, M.S., Lorence, R.M., Groene, W.S., Bamat, M.K.: Naturally oncolytic viruses. Curr. Opin. Mol. Ther. **8**, 314–321 (2006)
33. Vaha-Koskela, M.J., Heikkila, J.E., Hinkkanen, A.E.: Oncolytic viruses in cancer therapy. Cancer Lett. **254**, 178–216 (2007)
34. Wong, H.H., Lemoine, N.R., Wang, Y.: Oncolytic viruses for cancer therapy: overcoming the obstacles. Viruses **2**, 78–106 (2010)
35. Sun, W., Jiang, T., Lu, Y., Reiff, M. et al.: Cocoon-like self-degradable DNA nanoclew for anticancer drug delivery. J. Am. Chem. Soc. **136**, 14722–14725 (2014)
36. Bocharov, G., Ford, N.J., Edwards, J., Breining, T., et al.: A genetic-algorithm approach to simulating human immunodeficiency virus evolution reveals the strong impact of multiply infected cells and recombination. J. Gen. Virol. **86**, 3109–3118 (2005)
37. Sardanyés, J., Elena, S.F.: Quasispecies spatial models for RNA viruses with different replication modes and infection strategies. PLOS One **6**, 24884 (2011)
38. Ochoa, G.: Error thresholds in genetic algorithms. Evol. Comput. **14**(2), 157–182 (2006)
39. Vogelstein, B. Kinzler, K.W.: Cancer genes and the pathways they control. Nat. Med. **10**, 789–799 (2004)
40. Sardanyés, J., Elena, S.F., Solé, R.V.: Simple quasispecies models for the Survival-of-the-flattest effect: the role of space. J. Theor. Biol. **250**(3), 560–568

Guyer–Krumhansl Heat Conduction in Thermoreflectance Experiments

Matthew G. Hennessy and Tim G. Myers

Abstract Thermoreflectance experiments involve heating the surface of a solid using a high-frequency laser. The small length and time-scales associated with this rapid heating lead to the onset of heat conduction mechanisms that cannot be captured using Fourier's law. We propose a model for thermoreflectance experiments based on the Guyer–Krumhansl equation of heat conduction. We show that heat conduction occurs in the form of two distinct modes that are analogous to pressure and shear waves in linear viscoelastic materials. We present analytical solutions to the model that can be used to calculate the three-dimensional temperature and flux profiles in the heated solid as well as the phase difference between the laser and the surface temperature oscillations. Using the Laplace transform, we show how the solution can be extended to account for laser pulses with an arbitrary dependence on time.

1 Introduction

Nanotechnology is a rapidly growing interdisciplinary area, with breakthroughs having important implications in fields such as medicine, electronics and energy storage. One of the key issues with designing nanodevices is inefficient dissipation of heat, which can lead to performance degradation, melting and ultimately device failure [1, 2]. Theoretical and experimental studies have shown that there is a marked decrease in the thermal conductivity at the nanoscale [3–5], which prevents generated heat from being dissipated or exchanged with the surrounding environment. As the characteristic size of the system decreases, heat conduction becomes increasingly dominated by the infrequency of collisions between thermal

M. G. Hennessy (✉)
Mathematical Institute, University of Oxford, Oxford, UK
e-mail: hennessy@maths.ox.ac.uk

T. G. Myers
Centre de Recerca Matemàtica, Bellaterra, Barcelona, Spain
e-mail: tmyers@crm.cat

F. Font, T. G. Myers (eds.), *Multidisciplinary Mathematical Modelling*,
SEMA SIMAI Springer Series 11, https://doi.org/10.1007/978-3-030-64272-3_2

energy carriers known as phonons. The finite time between successive phonon collisions results in a ballistic transport process that effectively reduces the thermal conductivity of the system. Conversely, classical heat conduction is a diffusive process that occurs when the characteristic length and time-scales are much larger than the mean free path and mean free time of phonon collisions.

A number of experiments have been designed to access the extremely small length and time-scales associated with nanoscale heat transport. In thermoreflectance experiments, a high-frequency laser (called the pump beam) is used to heat the surface of a metallic film that is bonded to a sample of semiconducting material such as silicon or germanium [6, 7]. A secondary laser (called the probe beam) is used to measure the change in reflectance of the metal film. From these measurements, the surface temperature of the bilayer can be deduced and compared to theoretical predictions. Unsurprisingly, models based on Fourier's law which use the bulk value of the thermal conductivity are in poor agreement with experimental data as a consequence of the ballistic nature of nanoscale heat conduction not being accounted for. It is therefore common to interpret thermoreflectance experiments through the use of effective Fourier models (EFMs), which introduce features such as anisotropic thermal conductivity tensors [8] or frequency-dependent material properties [7] in order to match the experimental data. Although EFMs provide important insights into thermoreflectance experiments, they are bound to the assumption of diffusive transport and, as a result, can fail to provide consistent interpretations of the data. For example, fitting an EFM to data of the amplitude of oscillations in the surface temperature results in a thermal conductivity that monotonically increases with frequency; however, fitting to data of the phase of the oscillations leads to a conductivity that monotonically decreases with frequency [9].

The failure of Fourier's law to capture thermoreflectance data motivates the need for extended models that better capture the ballistic nature of nanoscale heat conduction. The first extension to Fourier's law was proposed by Cattaneo [10], who argued that a finite amount of time is required for a temperature gradient to produce a flux of thermal energy. By expanding the governing equations about a small delay time, the Maxwell–Cattaneo (MC) equation (or Maxwell–Cattaneo–Vernotte equation) can be obtained. When the MC equation is combined with the law of conservation of energy, the hyperbolic heat equation (HHE) is obtained, which captures the wave-like propagation of heat associated with ballistic transport. Although the HHE produces wave-like solutions, the ad hoc introduction of a delay time does little to explain the underlying physics of heat conduction. Guyer and Krumhansl [11, 12] later derived an extension to the MC equation directly from the linearised Boltzmann equation that explicitly accounts for the mean free time and mean free path of phonon collisions. From a theoretical point of view, the Guyer–Krumhansl (GK) equation is particularly attractive because it provides a link between kinetic and continuum models and is based on parameters that have a well-defined meaning in terms of the underlying physics of heat conduction. Moreover, the analogous form of the GK and Navier–Stokes equations enables nanoscale heat transport to be interpreted in terms of fluid mechanics, and this analogy has been used to explain the reduced thermal conductivity of nanosystems [13, 14].

The thermal response of a solid that is heated by a laser has been studied using non-Fourier conduction laws by several authors. Zhang et al. [15] considered a three-dimensional axisymmetric problem where the thermal flux was assumed to satisfy the MC equation. Maassen and Lundstrom [16] considered a one-dimensional situation and compared the results obtained from the MC equation to those from the lattice Boltzmann equation (LBE), showing that the two approaches are in excellent agreement. Kovács [17] also considered a one-dimensional geometry and modelled heat flow using the GK equation. Beardo et al. [9] studied three-dimensional heat conduction in thermoreflectance experiments using a model based on the GK equation which accounted for the thin metal film that is bonded to the surface of a semiconductor. They developed an analytical solution by assuming that the laser heating is harmonic in time.

In this chapter, we also model thermoreflectance experiments using the GK equation. We neglect the thin metal film used in experiments but allow for the laser heating to have an arbitrary dependence on time. We show that the full unsteady form of the thermal model is analogous to the Kelvin–Voigt model of a linear viscoelastic material. We use this analogy to decompose the solution of the thermal model into modes that mimic viscoelastic pressure and shear waves.

The outline of the chapter is as follows. In Sect. 2, we present the governing equations. In Sect. 3, the equations are related to those of linear viscoelasticity, and the solution is decomposed into wave-like modes. In Sect. 4, semi-analytical and analytical solutions are constructed when the laser heating is harmonic in time. In Sect. 5, we discuss how the previous solutions can be extended to account for arbitrary forms of the laser heating through the use of Laplace transforms. The chapter then concludes in Sect. 6.

2 Mathematical Model

We consider the heating of an isotropic solid sample with a high-frequency laser as shown in Fig. 1. Conservation of energy within the sample requires

$$c_v \dot{T} + \nabla \cdot \boldsymbol{q} = 0, \tag{1}$$

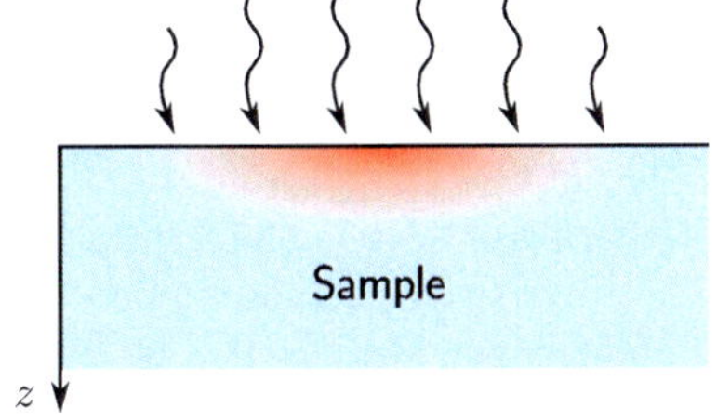

Fig. 1 A schematic diagram of a thermoreflectance experiment whereby the surface of a solid is heated by a high-frequency laser with a Gaussian intensity distribution

where c_v is the volumetric heat capacity (at constant volume) of the sample, T is the temperature difference relative to the initial and ambient temperature T_0 and $\boldsymbol{q} = (q_x, q_y, q_z)$ is the thermal flux. It is assumed that the optical extinction length is sufficiently small and that the absorption of radiation can be captured through a boundary condition rather than a volumetric heating term in (1). The thermal flux is assumed to obey the Guyer–Krumhansl (GK) equation

$$\tau_R \dot{\boldsymbol{q}} + \boldsymbol{q} = -k_{\text{bulk}} \nabla T + \ell^2 \left[\nabla^2 \boldsymbol{q} + 2\nabla(\nabla \cdot \boldsymbol{q}) \right], \tag{2}$$

where τ_R, k_{bulk} and ℓ are the relaxation time of resistive phonon collisions, bulk thermal conductivity and bulk mean free path (MFP) of phonons in the sample. Mathematically, the first term on the left-hand side of (2) accounts for memory effects and leads to the thermal flux depending on the time integral of the temperature gradient. The second and third terms on the right-hand side, i.e. those proportional to ℓ^2, represent non-local effects and lead to the flux depending on the spatial integral of the temperature gradient. The thermal conductivity and MFP can be written as $k_{\text{bulk}} = (1/3)c_v v_g^2 \tau_R$ and $\ell^2 = (1/5)v_g^2 \tau_R \tau_N$ with v_g being the speed of phonons in the material and τ_N the relaxation time of normal phonon collisions. Furthermore, we introduce the bulk thermal diffusivity $\kappa = k_{\text{bulk}}/c_v$.

The laser induces a thermal flux at the irradiated surface of the solid given by

$$\boldsymbol{q} \cdot \boldsymbol{e}_z = q_0 I(x, y) f(t), \quad z = 0, \tag{3}$$

where $\boldsymbol{e}_z$ is the unit vector normal to the surface with the coordinate z pointing into the sample, I describes the non-dimensional in-plane distribution of the laser intensity, x and y are horizontal coordinates centred at the point of irradiation, f describes the temporal aspects of the laser pulse and t is time. We will assume that the beam profile is Gaussian and given by

$$I(x, y) = \mathrm{e}^{-2(x^2+y^2)/r_b^2}, \tag{4}$$

where r_b is the $1/\mathrm{e}^2$ radius of the beam and q_0 represents the maximum intensity of the laser, which is related to the laser power P through the relation $q_0 = P/(\pi r_b^2)$.

Following previous works [18–20], slip-like boundary conditions for the tangential components of the thermal flux are specified,

$$q_x = Cl \frac{\partial q_x}{\partial z}, \quad q_y = Cl \frac{\partial q_y}{\partial z}; \quad z = 0. \tag{5}$$

The dimensionless parameter C encodes information about the nature of phonon scattering at the boundary, surface roughness and temperature effects [13, 20, 21]. Here, we make the assumption that phonons undergo diffusive scattering at the boundary and set $C = 1$.

Table 1 Temperature-dependent parameter values

Parameter [unit]	81 K	311 K
c_v [KJ m^{-3} K^{-1}]	466	1692
k_{bulk} [W m^{1} K^{-1}]	1260	150
τ [ps]	1002	42
ℓ [nm]	3127	185

We also impose far-field conditions of the form $T/T_0 \to 0$ and $q/q_0 \to 0$ as $z/L_F \to \infty$, where $L_F = (2\kappa/\omega)^{1/2}$ is the classical thermal penetration depth predicted with Fourier's law. We also assume that $T = 0$ and $\boldsymbol{q} = 0$ when $t = 0$.

The parameter values are based on the thermoreflectance experiments in Regner et al. [6, 7] involving c-Si. In these experiments, the radius of the pump beam was $r_b = 3.2\,\mu\text{m}$. When required, we assume that the power of the pump beam is $P = 1$ mW. The thickness of the layer (in the z direction) was $H = 525\,\mu\text{m}$. Since the thickness of the sample is much greater than the maximum thermal penetration depth, $H/L_F \gg 1$, we assume that the sample can be treated as infinitely thick. The temperature-dependent parameter values are listed in Table 1. The resistive relaxation time and MFP are taken from Beardo et al. [9].

3 Connections to Linear Viscoelasticity

Elimination of the temperature T from (2) shows that the thermal flux satisfies

$$\tau_R \ddot{\boldsymbol{q}} + \dot{\boldsymbol{q}} = \kappa \nabla(\nabla \cdot \boldsymbol{q}) + \ell^2 \left[\nabla^2 \dot{\boldsymbol{q}} + 2\nabla(\nabla \cdot \dot{\boldsymbol{q}}) \right]. \tag{6}$$

The bulk equation (6) and the boundary condition (3) show that the governing equations for the thermal flux are the same as those of a damped Kelvin–Voigt viscoelastic material with zero shear modulus that is subject to a fixed surface displacement. More specifically, the resistive relaxation time is analogous to the density, the thermal diffusivity is the Lamé parameter and the square of the MFP is the viscosity (shear and bulk).

Motivated by the analogy between nanoscale heat transfer and linear viscoelastic materials, we introduce the Helmholtz decomposition of the flux,

$$\boldsymbol{q}/q_0 = \nabla \Phi + \nabla \times \boldsymbol{A}, \tag{7}$$

in order to simplify the bulk equation (6). The scalar and vector potentials Φ and $\boldsymbol{A}$, respectively, can be shown to satisfy

$$\tau_R \ddot{\Phi} + \dot{\Phi} = \kappa \nabla^2 \Phi + 3\ell^2 \nabla^2 \dot{\Phi}, \tag{8a}$$

$$\tau_R \dot{\boldsymbol{A}} + \boldsymbol{A} = \ell^2 \nabla^2 \boldsymbol{A}, \tag{8b}$$

$$\nabla \cdot \boldsymbol{A} = 0. \tag{8c}$$

In the context of linear elasticity, the scalar and vector potentials capture pressure (P) and shear (S) waves, respectively. Both waves are non-dispersive; however, P-waves are longitudinal and irrotational, whereas S-waves are transverse and incompressible (i.e. volume conserving). In the context of heat flow, S-waves represent temperature-conserving thermal waves that propagate in orthogonal directions to the flux and do not influence the temperature. Only P-waves can generate a change in temperature, which, mathematically, can be seen by combining (1) and (7) to obtain $c_v\dot{T} = -q_0\nabla^2\Phi$. In the classical limit, $\omega\tau_R \to 0$ and $\ell/L_F \to 0$, we find that $\boldsymbol{A} = 0$, and thus the scalar potential simply becomes a multiple of the temperature, $\Phi = -(k_{\text{bulk}}/q_0)T$.

The problem for the vector potential can be simplified by introducing the 'streamfunction' ψ and writing $\boldsymbol{A} = \nabla \times (\psi\boldsymbol{e}_z)$. This form of $\boldsymbol{A}$ automatically fulfils the condition (8c). Moreover, the problem for ψ is given by

$$\nabla^2(\tau_R\dot{\psi} + \psi - \ell^2\nabla^2\psi) = 0. \tag{9}$$

To make further progress on the problem, we must specify the form of the heating function $f(t)$ in (3).

4 Time-Harmonic Heating

The laser pulses are first assumed to be harmonic in time, $f(t) = \mathrm{e}^{-i\omega t}$, where ω is the angular frequency of the laser. The problem can be simplified by first taking the two-dimensional (in-plane) Fourier transform of the governing equations and then seeking separable solutions of the form $\tilde{g}(\boldsymbol{k})\mathrm{e}^{-\lambda(\boldsymbol{k})z-i\omega t}$, where $\boldsymbol{k} = (k_x, k_y, 0)$ is a two-dimensional wave vector. The two-dimensional inverse Fourier transforms can be converted into one-dimensional Hankel transforms due to the axisymmetry of the problem, which can then be numerically evaluated using fast methods [22]. The scalar potential and streamfunction are given by

$$\Phi(r, z, t) = \int_0^\infty \tilde{\Phi}(k)\tilde{I}(k)\mathrm{e}^{-\lambda_p(k)z-i\omega t} J_0(kr)k\,\mathrm{d}k, \tag{10a}$$

$$\psi(r, z, t) = \int_0^\infty \tilde{\psi}(k)\tilde{I}(k)\mathrm{e}^{-\lambda_s(k)z-i\omega t} J_0(kr)k\,\mathrm{d}k, \tag{10b}$$

where $r = (x^2 + y^2)^{1/2}$, $k = |\boldsymbol{k}|$ and $\tilde{I}$ is the Hankel transform of the Gaussian beam profile in (4). The functions λ_p and λ_s are defined as

$$\lambda_p(k) = \left\{k^2 - \frac{2i(1 - i\omega\tau_R)}{L_F^2\left[1 - 6i(\ell/L_F)^2\right]}\right\}^{1/2}, \tag{11a}$$

$$\lambda_s(k) = \left\{k^2 + \ell^{-2}(1 - i\omega\tau_R)\right\}^{1/2}, \tag{11b}$$

with $\mathrm{Re}\{\lambda_s\} > 0$ and $\mathrm{Re}\{\lambda_p\} > 0$. The penetration depths of the thermal P- and S-modes are given by $L_\mathrm{P} = 1/\mathrm{Re}\{\lambda_p(0)\}$ and $L_\mathrm{S} = 1/\mathrm{Re}\{\lambda_s(0)\}$. The ratio ℓ/L_F defines an effective Knudsen number, characterising the relevance of non-local effects when the characteristic length scale of the system is L_F. The Fourier coefficients $\tilde{\Phi}$ and $\tilde{\psi}$ are obtained from the boundary conditions (3) and (5), resulting in

$$\tilde{\Phi}(k) = \frac{\lambda_s(1 + C\ell\lambda_s)}{k^2(1 + C\ell\lambda_p) - \lambda_s\lambda_p(1 + C\ell\lambda_s)}, \tag{12a}$$

$$\tilde{\psi}(k) = \frac{1 + C\ell\lambda_p}{k^2(1 + C\ell\lambda_p) - \lambda_s\lambda_p(1 + C\ell\lambda_s)}. \tag{12b}$$

The temperature is given by

$$T(r, z, t) = \int_0^\infty \tilde{T}(k)\tilde{I}(k)\mathrm{e}^{-\lambda_p(k)z - i\omega t} J_0(kr)k\,\mathrm{d}k, \tag{13a}$$

$$\tilde{T}(k) = -\frac{q_0}{k_\mathrm{bulk}}\left(\frac{1 - i\omega\tau_R}{1 - 6i(\ell/L_F)^2}\right)\tilde{\Phi}, \tag{13b}$$

which can be simplified to

$$T(r, z, t) = -\frac{q_0}{k_\mathrm{bulk}}\left(\frac{1 - i\omega\tau_R}{1 - 6i(\ell/L_F)^2}\right)\Phi(r, z, t). \tag{14}$$

The latter form provides a direct link between the temperature and the amplitude of P-waves in the case of GK conduction. In the case of Fourier conduction, we find that $\lambda_p = (k^2 - 2iL_F^{-2})^{1/2}$, $\tilde{\Phi} = -\lambda_p^{-1}$ and $\tilde{T} = -(q_0/k_\mathrm{bulk})\tilde{\Phi}$.

Figure 2 shows the two-dimensional temperature profiles at a given time t_* obtained using Fourier's law and the GK equation when the ambient temperature is $T_0 = 81\,\mathrm{K}$. We take $f = 10^8\,\mathrm{Hz}$, which leads to $L_F \simeq 2.9\,\mu\mathrm{m}$, $\omega\tau_R \simeq 0.63$ and $\ell/L_F \simeq 1.07$. The time t_* is chosen so that the real part of the temperature solution (13) evaluated at the origin is maximised. Mathematically, t_* is the solution to $\max_{0 \leq t \leq 2\pi/\omega} \mathrm{Re}\{T(0, 0, t)\}$. In the case of Fourier's law, the isotherms are roughly elliptical, indicating that the temperature monotonically decreases with distance from the origin. When the GK equation is used, the isotherms become more complex and indicate that the temperature is non-monotonic in the radial and axial directions. Furthermore, there is a region near the illuminated surface ($z = 0$) where the temperature becomes negative for intermediate values of r/L_F. This region is seen more clearly in Fig. 3, which plots the surface temperature profiles at various times. The profiles in the case of GK conduction correspond to standing waves that have been spatially damped due to the inclusion of non-local effects (or 'thermal viscosity'). Interestingly, this dampening leads to a localisation of the surface heating; the temperature increase is more confined to where the laser hits the surface (note that $r_b/L_F \simeq 0.77$) compared to the Fourier case. As a result of

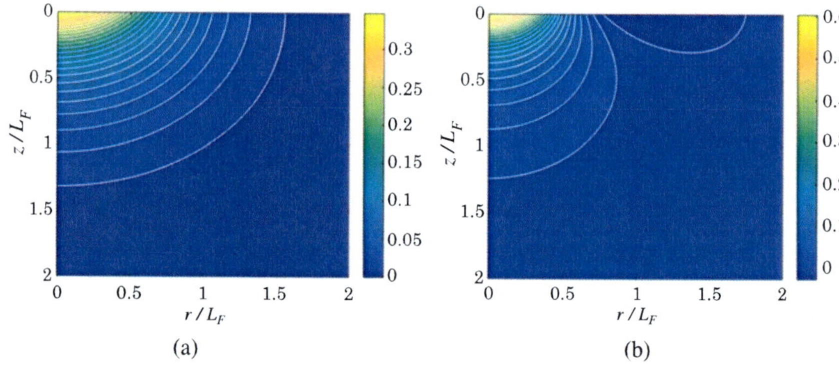

Fig. 2 Temperature profiles obtained using Fourier's law (**a**) and the GK equation (**b**) when $T_0 = 81\,\text{K}$ and $f = 10^8\,\text{Hz}$. The temperature has been normalised by $q_0 L_F/k_{\text{bulk}}$. (**a**) Fourier conduction. (**b**) GK conduction

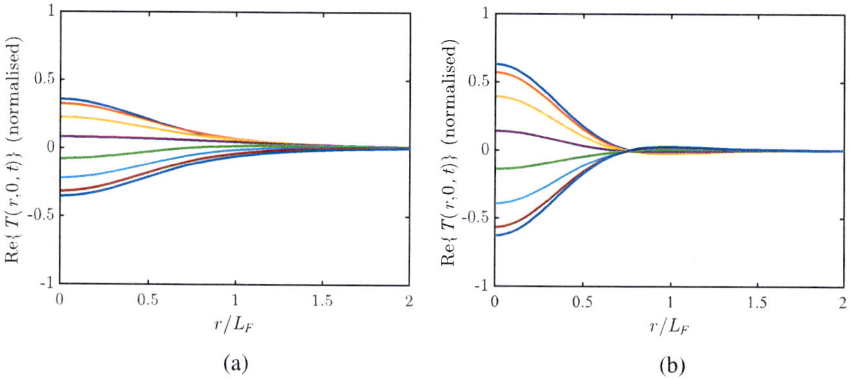

Fig. 3 Time evolution of the surface temperature obtained using Fourier's law (**a**) and the GK equation (**b**) when $T_0 = 81\,\text{K}$ and $f = 10^8\,\text{Hz}$. The temperature has been normalised by $q_0 L_F/k_{\text{bulk}}$. The solutions are shown at times $t = t_* + n\pi/(7\omega)$, where $n = 0, 1, \ldots, 7$. The definition of t_* can be found in the main body of the chapter. (**a**) Fourier conduction. (**b**) GK conduction

this localisation, the amplitude of the surface temperature oscillations is greater in the case of GK conduction.

In thermoreflectance experiments, a probe beam is used to measure the change in reflectivity of the sample. This enables a spatially integrated form of the surface temperature to be inferred. Following Cahill [23], the surface temperature measured from a probe beam with the same profile as the pump beam can be calculated as

$$\Delta T_m = \frac{r_b^2}{4}\int_0^\infty \tilde{T}(k)\mathrm{e}^{-r_b^2 k^2/4} k\,\mathrm{d}k. \tag{15}$$

Another key experimentally measurable quantity is the phase difference between the oscillations of the surface temperature and the radiation provided by the pump beam. This phase difference can be calculated from (15) as $\varphi = -\arg\{\Delta T_m\}$.

To study the dependence of the measured temperature ΔT_m and the phase difference φ on frequency, we first assume that the MFP of phonons, ℓ, is smaller than the beam radius, r_b, which is true when the ambient temperature is $T_0 = 311\,\mathrm{K}$. For small angular frequencies that satisfy $\omega \ll \kappa/r_b^2$ with $\omega\tau_R \ll 1$ and $\ell/L_F \ll 1$, the measured temperature coincides with the low-frequency Fourier-based prediction,

$$\Delta T_m \simeq \Delta T_{\mathrm{LF}} = \frac{1}{2\pi^{1/2}} \frac{P}{k_{\mathrm{bulk}} r_b}. \tag{16}$$

In this case, the surface temperature is in phase with the pump beam, $\varphi \simeq 0$, and heat is conducted in all the three dimensions.

For larger frequencies that satisfy $\omega \gg \kappa/r_b^2$, the classical penetration depth L_F is much smaller than the beam radius r_b, and heat conduction primarily occurs in the axial direction. In this case, the measured temperature can be approximated by the solution obtained from a one-dimensional model,

$$\Delta T_m \simeq \Delta T_{\mathrm{HF}} \left(\frac{1 - i\omega\tau_R}{1 - 6i(\ell/L_F)^2} \right)^{1/2}, \quad \Delta T_{\mathrm{HF}} = \frac{P}{2\pi r_b^2} \left(\frac{i}{k_{\mathrm{bulk}} \rho c_v \omega} \right)^{1/2}. \tag{17}$$

The prefactor ΔT_{HF} corresponds to the high-frequency limit in the case of Fourier conduction. The corresponding phase shift is given by

$$\varphi \simeq -\frac{\pi}{4} + \frac{1}{2} \arctan(\omega\tau_R) - \frac{1}{2} \arctan\left(\frac{6\ell^2}{L_F^2} \right). \tag{18}$$

The first component, $\varphi_{\mathrm{HF}} = -\arg\{\Delta T_{\mathrm{HF}}\} = -\pi/4$, is a frequency-independent contribution from Fourier conduction. The second and third contributions are attributed to resistive and normal phonon scattering modes, respectively. For values of the angular frequency given by $\omega \gg \tau_R^{-1}$ and $\omega \gg \kappa/\ell^2$, the one-dimensional approximation given by (17) simplifies even further to $\Delta T_m/\Delta T_{\mathrm{HF}} \sim (\tau_R\kappa)^{1/2}/\ell$, which indicates that the measured temperature retains a scaling that is proportional to $\omega^{-1/2}$. Moreover, the non-Fourier contributions to the phase shift (18) cancel out in the high-frequency limit, and the Fourier result $\varphi \simeq -\pi/4$ is recovered. In the special case when $\tau_R\kappa/\ell^2 = 1$, no distinction can be made between Fourier and GK conduction, a phenomenon that is sometimes referred to as Fourier resonance [24, 25]. Differences in the magnitude of $\tau_R\kappa/\ell^2$ lead to distinct behaviour in the regime of intermediate frequencies. When $\tau_R \ll \ell^2/\kappa$, which is the case for the parameters in Table 1, there is a range of frequencies given by $\kappa/\ell^2 \ll \omega \ll \tau_R^{-1}$ where the measured temperature $|\Delta T_m|$ scales like ω^{-1} and thus

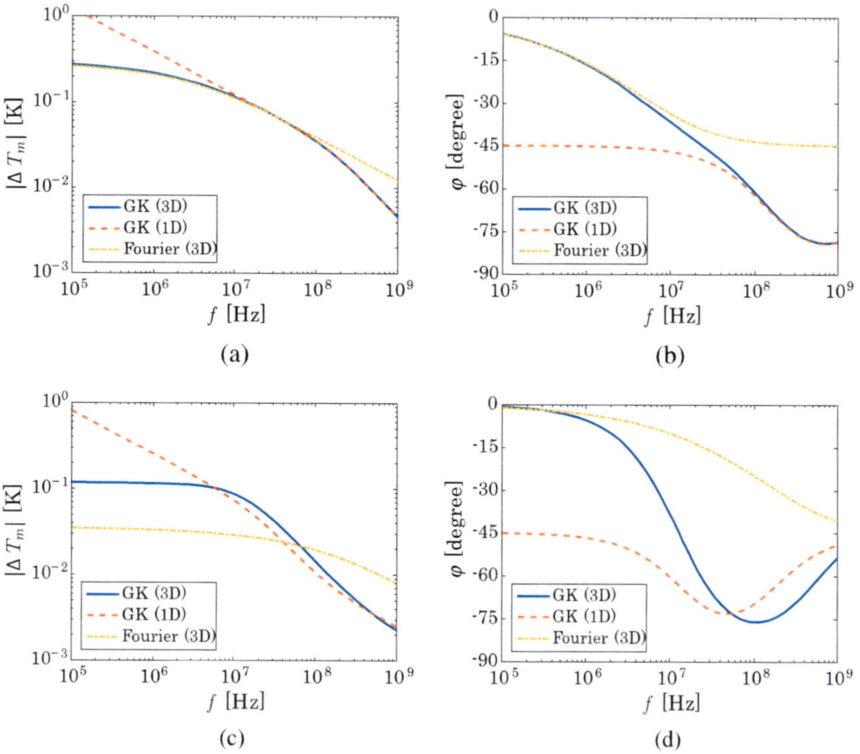

Fig. 4 The frequency dependence of the amplitude $|\Delta T_m|$ and the phase difference φ of surface temperature oscillations measured with a probe beam computed using Fourier's law, the GK equation and the one-dimensional GK equation. (**a**) $T_0 = 311$ K. (**b**) $T_0 = 311$ K. (**c**) $T_0 = 81$ K. (**d**) $T_0 = 81$ K

decreases faster with frequency than predicted by Fourier's law. Furthermore, the phase difference decreases beyond the Fourier value and approaches $\varphi \simeq -\pi/2$. This is shown more clearly in Fig. 4a, b, in which $|\Delta T_m|$ and φ are plotted as a function of frequency. Conversely, if $\tau_R \gg \ell^2/\kappa$, then there is a range of frequencies given by $\tau_R^{-1} \ll \omega \ll \kappa/\ell^2$ where $|\Delta T_m|$ exhibits no dependence on the frequency, leading to a higher temperature than predicted by Fourier's law, consistent with the numerical results obtained from the HHE and phonon LBE [16]. In this case, the phase difference begins to increase and approach $\varphi = 0$.

In Fig. 4a, b, we also compare the amplitude of the measured temperature $|\Delta T_m|$ and the phase difference obtained from Fourier's law, the GK equation and the one-dimensional GK solutions given by (17) and (18) when $T_0 = 311$ K. From these figures, we can deduce the existence of three distinct regimes of heat conduction at this temperature. At low frequencies, heat conduction is three-dimensional and diffusive, that is, it can be captured using Fourier's law. As the frequency is increased, non-local effects become relevant, and the three-dimensional

GK conduction begins to occur. However, for very high frequencies, heat conduction becomes one-dimensional. The plateau in the phase difference that occurs when $f = 10^9$ Hz indicates that resistive phonon scattering is also beginning to become relevant.

Figure 4c, d illustrates the frequency dependence of $|\Delta T_m|$ and φ when $T_0 = 81$ K. In this case, the phonon MFP ℓ is roughly the same size as the beam radius. As a result, there are substantial deviations between the Fourier and the GK temperature amplitudes even at low frequencies. Moreover, the phase difference reveals that heat conduction remains three-dimensional across the entire frequency band.

5 Arbitrary Heating Functions

Using Laplace transforms, it is possible to construct an analytical solution for the temperature for arbitrary forms of the heating function $f(t)$. We let $\bar{f}(s) = \mathcal{L}\{f(t)\}$ denote the Laplace transform of f. Then, the solution for the temperature is given by $T(r, z, t) = \mathcal{L}^{-1}\{\bar{T}(r, z; s)\}$, where

$$\bar{T}(r, z; s) = \int_0^\infty \tilde{T}(k; s)\tilde{I}(k)\mathrm{e}^{-\lambda_p(k;s)z}\bar{f}(s)J_0(kr)k\,\mathrm{d}k. \quad (19)$$

Here, the functions $\tilde{T}(k; s)$ and $\lambda_p(k; s)$ are given by (11a) and (13b) with ω replaced by is.

In Fig. 5, we compare the spatio-temporal evolution of the surface temperature profile $T(r, 0, t)$ by numerically inverting the Hankel and the Laplace transforms in the case of Fourier and GK conduction. We have taken $f(t) = \exp(-t/\tau)$, where $\tau = 10^{-7}$ s and $T_0 = 81$ K. Similar to when the heating function was harmonic in time, in this case, we see that GK conduction also leads to a localisation of the

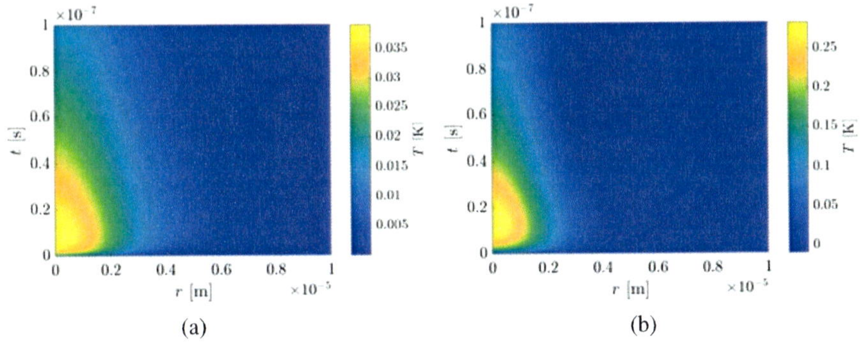

Fig. 5 Surface temperature profiles induced by a laser pulse with an amplitude that varies in time according to $f(t) = \exp(-t/\tau)$ with $\tau = 10^{-7}$ s. The ambient temperature is $T_0 = 81$ K. (**a**) Fourier conduction. (**b**) GK conduction

temperature profile. Consequently, the maximum temperature induced by the pulse is an order of magnitude greater compared to the Fourier case.

6 Conclusions

We have used the GK equation to study heat conduction in thermoreflectance experiments. By exploiting the analogy between nanoscale heat transfer and viscoelastic wave propagation, analytical solutions to the problem were constructed. In the case of time-harmonic heating, we found that there were three main regimes that can occur. For low frequencies, three-dimensional Fourier conduction was recovered. For intermediate frequencies, three-dimensional GK conduction occurs. Finally, for high frequencies, GK heat conduction becomes one-dimensional. The results also revealed that non-local effects due to normal phonon scattering introduce a thermal viscosity into the model. The additional resistance to heat flow leads to a localisation of the surface heating and higher temperatures compared to the Fourier case. However, the measurements of the surface temperature obtained using a probe beam, which are essentially the integrated product of the surface temperature and the beam profile, can result in values that are smaller than those predicted by Fourier's law. Thus, these measurements can miss the thermal localisation that occurs due to non-Fourier conduction.

We considered radially symmetric systems involving pump and probe beams with concentric circular cross sections. This allowed the solution to be written in terms of Hankel transforms. However, by using two-dimensional Fourier transforms, the solutions presented here could easily be adapted to handle a range of non-symmetric situations, such as those involving elliptical beams [26] or pump and probe beams with an offset [27]. These can provide deeper insights into the heat conduction mechanisms that occur at the nanoscale and further means of validating the GK equation or other non-Fourier conduction laws.

Acknowledgments This project has received funding from the European Union's Horizon 2020 research and innovation programme under grant agreement no. 707658. MC acknowledges that the research leading to these results has received funding from 'la Caixa' Foundation. TM acknowledges financial support from the Ministerio de Ciencia e Innovación grant MTM2017-82317-P. The authors have been partially funded by the CERCA Programme of the Generalitat de Catalunya.

References

1. Cahill, D.G., Ford, W.K., Goodson, K.E., Mahan, G.D., Majumdar, A., Maris, H.J., Merlin, R., Phillpot, S.R.: Nanoscale thermal transport. J. Appl. Phys. **93**(2), 793–818 (2003)
2. Cahill, D.G., Braun, P.V., Chen, G., Clarke, D.R., Fan, S., Goodson, K.E., Keblinski, P., King, W.P., Mahan, G.D., Majumdar, A. and Maris, H.J., Phillpot, S.R., Pop, E., Shi, L.: Nanoscale thermal transport. II. 2003–2012. Appl. Phys. Rev. **1**(1), 011305 (2014)

3. Ashenghi, M., Leung, Y.K., Wong, S.S., Goodson, K.E.: Phonon-boundary scattering in thin silicon layers. Appl. Phys. Lett. **71**(13), 1798–1800 (1997)
4. Li, D., Wu, Y., Kim, P., Shi, L., Yang, P., Majumdar, A.: Thermal conductivity of individual silicon nanowires. Appl. Phys. Lett. **83**(14), 2934–2936 (2003)
5. Liu, W., Asheghi, M.: Phonon-boundary scattering in ultrathin single-crystal silicon layers. Appl. Phys. Lett. **84**(19), 3819–3821 (2004)
6. Regner, K.T., Majumdar, S., Malen, J.A.: Instrumentation of broadband frequency domain thermoreflectance for measuring thermal conductivity accumulation functions. Rev. Sci. Instrum. **84**(6), 064901 (2013)
7. Regner, K.T., Sellan, D.P., Su, Z., Amon, C.H., McGaughey, A.J.H., Malen, J.A.: Broadband phonon mean free path contributions to thermal conductivity measured using frequency domain thermoreflectance. Nat. Commun. **4**, 1640 (2013)
8. Wilson, R.B., Cahill, D.G.: Anisotropic failure of Fourier theory in time-domain thermoreflectance experiments. Nat. Commun. **5**, 5075 (2014)
9. Beardo, A., Hennessy, M.G., Sendra, L., Camacho, J., Myers, T.G., Bafaluy, J., Alvarez, F.X.: Phonon hydrodynamics in frequency-domain thermoreflectance experiments. Phys. Rev. B **101**(7), 075303 (2020)
10. Cattaneo, C.: A form of heat conduction equation which eliminates the paradox of instantaneous propagation. Compte Rendus **247**(4), 431–433 (1958)
11. Guyer, R.A., Krumhansl, J.A.: Solution of the linearized phonon Boltzmann equation. Phys. Rev. **148**(2), 766 (1966)
12. Guyer, R.A., Krumhansl, J.A.: Thermal conductivity, second sound, and phonon hydrodynamic phenomena in nonmetallic crystals. Phys. Rev. **148**(2), 778 (1966)
13. Alvarez, F.X., Jou, D., Sellitto, A.: Phonon hydrodynamics and phonon-boundary scattering in nanosystems. J. Appl. Phys. **105**(1), 014317 (2009)
14. Jou, D., Casas-Vázquez, J., Lebon, G.: Extended Irreversible Thermodynamics, 4th edn. Springer, Berlin (2010)
15. Zhang, L., Shang, X.: Analytical solution to non-Fourier heat conduction as a laser beam irradiating on local surface of a semi-infinite medium. Int. J. Heat Mass Transf. **85**, 772–780 (2015)
16. Maassen, J., Lundstrom, M.: Modeling ballistic effects in frequency-dependent transient thermal transport using diffusion equations. J. Appl. Phys. **119**(9), 095102 (2016)
17. Kovács, R.: Analytic solution of Guyer-Krumhansl equation for laser flash experiments. Int. J. Heat Mass Transf. **127**, 631–636 (2018)
18. Calvo-Schwarzwälder, M., Hennessy, M.G., Torres, P., Myers, T.G., Alvarez, F.X.: A slip-based model for the size-dependent effective thermal conductivity of nanowires. Int. Commun. Heat Mass Transf. **91**, 57–63 (2018)
19. Calvo-Schwarzwälder, M., Hennessy, M.G., Torres, P., Myers, T.G., Alvarez, F.X.: Effective thermal conductivity of rectangular nanowires based on phonon hydrodynamics. Int. J. Heat Mass Transf. **126**, 1120–1128 (2018)
20. Sellitto, A., Alvarez, F.X., Jou, D.: Temperature dependence of boundary conditions in phonon hydrodynamics of smooth and rough nanowires. J. Appl. Phys. **107**(11), 114312 (2010)
21. Zhu, C.-Y., You, W., Li, Z.-Y.: Nonlocal effects and slip heat flow in nanolayers. Sci. Rep. **7**, 9568 (2017)
22. Chouinard, U., Baddour, N.: Matlab code for the discrete Hankel transform. J. Open Res. Softw. **5**(1), 4 (2017)
23. Cahill, D.G.: Analysis of heat flow in layered structures for time-domain thermoreflectance. Rev. Sci. Instrum. **75**(12), 5119–5122 (2004)
24. Both, S., Czél, B., Fülöp, T., Gróf, G., Gyenis, Á., Kovács, R., Ván, P., Verhás, J.: Deviation from the Fourier law in room-temperature heat pulse experiments. J. Non-Equilib. Thermodyn. **41**(1), 41–48 (2016)
25. Ván, P., Berezovski, A., Fülöp, T., Gróf, Gy., Kovács, R., Lovas, Á., Verhás, J.: Guyer-Krumhansl–type heat conduction at room temperature. Europhys. Lett. **118**(5), 50005 (2017)

26. Jiang, P., Qian, X., Yang, R.: A new elliptical-beam method based on time-domain thermoreflectance (TDTR) to measure the in-plane anisotropic thermal conductivity and its comparison with the beam-offset method. Rev. Sci. Instrum. **89**(9), 094902 (2018)
27. Feser, J.P., Cahill, D.G.: Probing anisotropic heat transport using time-domain thermoreflectance with offset laser spots. Rev. Sci. Instrum. **83**(10), 104901 (2012)

A Mathematical Model of Carbon Capture by Adsorption

Francesc Font, Tim G. Myers, and Matthew G. Hennessy

Abstract We present a model to describe the capture of carbon by an adsorbing porous material occupying a circular cross-sectional column. The model consists of an advection–reaction–diffusion equation for the gas concentration coupled to a simple kinetic equation describing gas adsorption on the pores. It is applicable to isothermal and isobaric gas transport with adsorption. The equations are defined in a domain with a free boundary. We obtain asymptotic solutions for large and small times and find good agreement with the numerical simulations. Our solutions show qualitative agreement with an experiment of carbon capture by adsorption in the literature.

1 Introduction

Since the 1950s, the presence of carbon in the atmosphere has increased by 36% such that current CO_2 levels are the highest in the last 800,000 years [1]. A recent study indicates that the major contribution to the flux of carbon into the atmosphere comes from human activity, mainly from burning fossil fuels, and is larger than, for instance, the carbon flux from volcanic and tectonic sources [2]. Large perturbations of the carbon cycle caused by large and fast carbon (particularly in the form of CO_2) influxes into the atmosphere have been the cause of global warming and dangerous alterations to the Earth's ecosystems in the past, which has even led to mass extinctions [2, 3]. It is therefore urgent that humankind takes measures to reduce CO_2 emissions.

A promising strategy to reduce carbon emissions consists of capturing the carbon directly from the exhaust of large CO_2 emission sources such as power plants or

F. Font (✉) · T. G. Myers
Centre de Recerca Matemàtica, Bellaterra, Barcelona, Spain
e-mail: ffont@crm.cat; tmyers@crm.cat

M. G. Hennessy
Mathematical Institute, University of Oxford, Oxford, UK
e-mail: hennessy@maths.ox.ac.uk

F. Font, T. G. Myers (eds.), *Multidisciplinary Mathematical Modelling*,
SEMA SIMAI Springer Series 11, https://doi.org/10.1007/978-3-030-64272-3_3

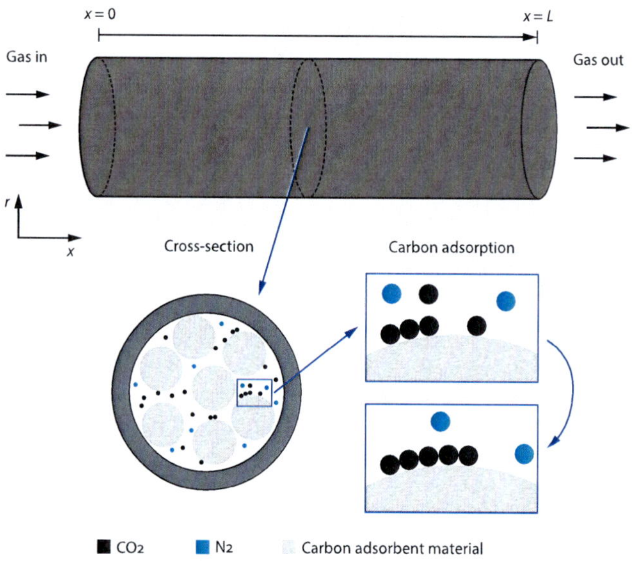

Fig. 1 A typical cylindrical column as used in experiments for carbon capture by adsorption. The white region in the cross-sectional panel represents the pores where the gas mixture travels. The CO_2 molecules are adsorbed on the pore surface of the carbon adsorbent material (represented by the large spheres), while the N_2 follows its course until the end of the column

cement manufacturing facilities. The most common methods to capture carbon are adsorption, absorption, separation by membranes and cryogenic separation [4–7]. In this work, we will present and analyse a mathematical model describing carbon capture by adsorption in a typical experimental set-up.

The process of carbon capture by adsorption involves a gas forced to circulate through a column packed with carbon adsorbent material (see Fig. 1). In the literature, the mathematical models describing the process of interest typically involve heat equations for the temperature of the gas and the adsorbent material and an advection–diffusion equation for the gas concentration [8–18]. It is standard for the advection–diffusion equation to contain a mass sink that accounts for the mass loss due to adsorption. In the heat equation (for the adsorbent material), the adsorption contributes through a source term due to the exothermic reaction. However, published numerical results typically show low temperature rises (of the order of 5K). In the current work, we will develop analytical and numerical solutions of the model describing carbon capture by adsorption assuming isothermal conditions, as formulated in [18] .

In the next section, we present and nondimensionalise the mathematical model. In Sects. 3 and 4, we derive asymptotic solutions for large and small times, respectively. In Sect. 5, we present a numerical strategy to solve the model. In

Sect. 6, we present and discuss the asymptotic and numerical solutions and compare them with experimental data from the literature. Finally, in Sect. 7, we draw our conclusions.

2 Mathematical Model

The typical experiments for carbon capture by adsorption involve a gas mixture, typically CO_2–N_2, forced through a cylindrical column that contains a CO_2 adsorbing porous material. As the gas passes through the pores, the CO_2 attaches to the surfaces, thereby capturing part of the carbon, while the rest of the gas components pass through. We illustrate this process in Fig. 1.

Assuming radial symmetry, isothermal, isobaric conditions, a constant gas velocity, and averaging through a circular cross section of the column, an equation for the averaged concentration in the axial direction was obtained

$$\frac{\partial c^*}{\partial t^*} + u\frac{\partial c^*}{\partial x^*} = D\frac{\partial^2 c^*}{\partial x^{*2}} - \frac{1-\epsilon}{\epsilon}\rho_q\frac{\partial q^*}{\partial t^*}, \tag{1}$$

where c^* and q^* are the averaged CO_2 concentration and amount adsorbed, respectively, and u is the interstitial velocity of the gas. The averaged quantities c^* and q^* are functions of the position in the axial direction (x^*) and time (t^*). The material parameters in (1) are described in Table 1.

The mass sink in (1) results from the integration over the radial component r^* during the averaging process and accounts for the mass loss due to CO_2 adsorption [18]. The sink rate $\frac{\partial q^*}{\partial t^*}$ can be modelled as the first-order kinetic reaction

$$\frac{\partial q^*}{\partial t^*} = k(q_e - q^*), \tag{2}$$

Table 1 Typical parameter values for CO_2 adsorption [15, 19, 20]

	Symbol	Value	Dimension
Initial concentration (CO_2)	c_0	6.03	mol/m^3
Adsorption saturation (CO_2)	q_e	1.57	mol/kg
Bed void fraction	ϵ	0.56	–
Density of adsorbed CO_2	ρ_q	330	kg/m^3
Axial diffusion coefficient	D	1.44×10^{-5}	m^2/s
Bed length	L	0.2	m
Interstitial velocity	u	0.02	m/s
Adsorption rate constant (CO_2)	k	0.014	s^{-1}

where q_e represents the maximum amount of adsorbate in the porous material and k is the adsorption rate. From now on, the averaged quantities c^* and q^* will be referred to simply as *concentration* and *adsorption*.

The boundary and initial conditions are as follows. At the inlet, we have the Danckwerts-type condition

$$uc_0 = \left(uc^* - D\frac{\partial c^*}{\partial x^*} \right)\bigg|_{x^*=0} . \tag{3}$$

At the outlet $x^* = L$, it is standard to assume that $c^*(L^-, t^*) \approx c^*(L^+, t^*)$, and so

$$\frac{\partial c^*}{\partial x^*}\bigg|_{x^*=L} = 0 . \tag{4}$$

However, this really only applies when CO_2 has reached the outlet, which can take a considerable amount of time. In practice, there is a moving boundary, beyond which the concentration is zero. So instead, we impose

$$\frac{\partial c^*}{\partial x^*}\bigg|_{x^*=s(t)} = c^*(s^*(t^*), t^*) = 0 , \tag{5}$$

where the free boundary is located at $x = s(t)$. Note that we have replaced the condition (4) by two conditions; the additional one is required to determine $s(t)$. Finally, the system is closed with the initial conditions

$$c^*(x^*, 0) = 0 , \qquad q^*(x^*, 0) = 0 , \qquad s^*(0) = 0 . \tag{6}$$

2.1 Nondimensional Model

To nondimensionalise our model, we take the parameter values from Table 1, which correspond to the CO_2 capture experiments in [15] (apart from the values of ρ_q and D, which are taken from [19] and [20], respectively).

We now introduce the dimensionless variables

$$c = \frac{c^*}{\mathcal{C}}, \qquad x = \frac{x^*}{\mathcal{L}}, \qquad t = \frac{t^*}{\mathcal{T}}, \qquad q = \frac{q^*}{\mathcal{Q}} \tag{7}$$

into the governing equations (1)–(2). Given that our primary interest is the removal of CO_2, we scale the concentration with the initial value of this component, so $\mathcal{C} = c_0$. The adsorption will then be scaled with its saturation value, $\mathcal{Q} = q_e$. The relevant time-scale at which adsorption will occur is controlled by the reaction rate in (2), so we choose $\tau = k^{-1} \approx 73.53$ s. The length scale can be chosen from the advection or diffusion terms in (1). Unless the gas velocity is extremely low, advection dominates over diffusion, so we choose to balance the advection term with the sink term and obtain $\mathcal{L} = \epsilon\, u\, c_0/(1-\epsilon)k\rho_q q_e \approx 0.02\,\text{m}$.

The resulting nondimensional governing equations are

$$\delta\frac{\partial c}{\partial t} + \frac{\partial c}{\partial x} = Pe^{-1}\frac{\partial^2 c}{\partial x^2} - \frac{\partial q}{\partial t}\,, \tag{8}$$

$$\frac{\partial q}{\partial t} = 1 - q\,, \tag{9}$$

where the dimensionless parameters are given by

$$\delta = \frac{kL}{u} = 0.01\,, \qquad Pe^{-1} = \frac{D}{L\,u} = 0.04\,. \tag{10}$$

Finally, the boundary conditions are

$$1 = \left(c - Pe^{-1}\frac{\partial c}{\partial x}\right)\bigg|_{x=0}\,, \qquad \frac{\partial c}{\partial x}\bigg|_{x=s(t)} = c(s(t), t) = 0\,, \tag{11}$$

and the problem is closed with the initial conditions $c(x, 0) = q(x, 0) = s(0) = 0$.

3 Travelling Wave Solution

If we define a new variable $\eta = x - s(t)$, which has its origin at the reaction front, then equations (8)–(9) become

$$-\delta\frac{ds}{dt}\frac{\partial c}{\partial \eta} + \frac{\partial c}{\partial \eta} = Pe^{-1}\frac{\partial^2 c}{\partial \eta^2} + \frac{ds}{dt}\frac{\partial q}{\partial \eta}\,, \tag{12}$$

$$-\frac{ds}{dt}\frac{\partial q}{\partial \eta} = 1 - q\,. \tag{13}$$

A self-similar form will be found if this equation depends on the single variable η, which means that there can be no time dependence. Consequently, we write $s_t = v =$ constant. The travelling wave form holds away from the initial time, so we make the approximation that the condition at $x = 0$, from (11), may be applied far behind the front, i.e. $\eta = -s(t)$ is large and negative, so we write

$$1 = \left(c - Pe^{-1}\frac{\partial c}{\partial \eta}\right)\bigg|_{\eta\to-\infty}\,, \tag{14}$$

and at $\eta = 0$, we have

$$\left.\frac{\partial c}{\partial x}\right|_{\eta=0} = c(0) = 0\,. \tag{15}$$

Integrating equation (12) and applying the boundary conditions (15), we obtain

$$(1 - \delta v)c = Pe^{-1}\frac{\partial c}{\partial \eta} + vq\;. \tag{16}$$

Now, by substituting (13) into (12) and subtracting (16) from the resulting expression, the variable q can be eliminated. This leads to the following equation for concentration:

$$vPe^{-1}\frac{\partial^2 c}{\partial \eta^2} - \left(Pe^{-1} + v - \delta v^2\right)\frac{\partial c}{\partial \eta} + (1 - \delta v)c = v\,, \tag{17}$$

which can be solved analytically

$$c = Ae^{\eta/v} + Be^{Pe(1-\delta v)\eta} + \frac{v}{1 - \delta v}\,. \tag{18}$$

The boundary conditions (15) give $A = -v^2/(v(1 - \delta v) - Pe^{-1})$ and $B = Pe^{-1}v/(1 - \delta v)(v(1 - \delta v) - Pe^{-1})$. Then, using (14), we find $v = 1/(1 + \delta)$. Finally, we obtain

$$c(x, t) = 1 + \frac{1}{v^2 Pe - 1}\left(e^{vPe\eta} - v^2 Pe\, e^{\eta/v}\right) \tag{19}$$

$$q(x, t) = 1 - e^{\eta/v}\,. \tag{20}$$

The variable η is fully defined when we have an expression for $s(t)$. Since we know that the velocity is constant, $v = 1/(1 + \delta)$, we may immediately write

$$s(t) = vt + s_0\,. \tag{21}$$

The constant of integration s_0 cannot be determined from the initial condition (since the solution does not hold for $t = 0$). It may be inferred using experimental data, for example, from the measurements of the outlet gas or from the numerical solution.

Since $\eta \le 0$ and $Pe \gg 1$, the first exponential is negligible, and we obtain

$$c(x, t) \approx 1 - \frac{v^2 Pe}{v^2 Pe - 1}e^{\eta/v}\,. \tag{22}$$

We can reduce this expression further by noting that for Pe large, we may write the concentration in series form

$$c(x,t) = 1 - \left(1 + \frac{1}{v^2 Pe} + \frac{1}{(v^2 Pe)^2} \cdots \right) e^{\eta/v} = 1 - \alpha e^{\eta/v}, \tag{23}$$

where we denote the series by α. Since $v^2 Pe \approx 15 \gg 1$, this series converges rapidly. If all terms involving $v^2 Pe$ are neglected, then the concentration is identical to the adsorption. So, we expect c and q to be very similar.

4 Asymptotic Solution for Small Times

To carry out a small time analysis, we change to a short time-scale τ such that $t = \varepsilon\tau$, where $\varepsilon \ll 1$. At small times, the amount adsorbed is also small, and we may write $q = \varepsilon Q$, and then

$$\frac{\partial Q}{\partial \tau} = 1 - \varepsilon Q, \tag{24}$$

subject to $Q(\tau_s(x)) = 0$. This suggests an asymptotic expansion in powers of ε. To first order, the appropriate solution is

$$Q = \tau - \tau_s - \varepsilon \frac{(\tau - \tau_s)^2}{2} + \cdots \tag{25}$$

At sufficiently small times, at leading order, the adsorption rate $\partial Q/\partial\tau = 1$ is independent of x. This is important for the concentration calculation.

The concentration now satisfies

$$\frac{\delta}{\varepsilon}\frac{\partial c}{\partial \tau} + \frac{\partial c}{\partial x} = Pe^{-1}\frac{\partial^2 c}{\partial x^2} - 1. \tag{26}$$

For the current problem, $\delta \approx 0.015$. In order to neglect the time dependence (so making the analysis tractable), we must then choose $\delta \ll \varepsilon \ll 1$, for example, by setting $\varepsilon = \sqrt{\delta}$. Then, for $x \in [0, s]$, the leading order concentration satisfies

$$\frac{\partial c}{\partial x} = Pe^{-1}\frac{\partial^2 c}{\partial x^2} - 1. \tag{27}$$

This has solution

$$c = A + Be^{Pex} - x. \tag{28}$$

Applying the leading order boundary conditions at $x = 0$ and $x = s$

$$1 = \left(c - Pe^{-1} \frac{\partial c}{\partial x} \right) \bigg|_{x=0} \quad , \quad c(s, \tau) = \frac{\partial c}{\partial x} \bigg|_{x=s} = 0 \tag{29}$$

results in $A = 1 - Pe^{-1}$, $B = Pe^{-1} e^{-Pes}$ and $s = 1$, and hence

$$c = 1 - Pe^{-1} + Pe^{-1} e^{-Pe(1-x)} - x \,. \tag{30}$$

This is the form of the concentration that varies between its inlet value and zero at the point $x = s$. It is independent of time, as is the position $s = 1$.

The conclusion that to leading order, for sufficiently small times, $s = 1$ may appear strange. In fact, over an even shorter time-scale, of order δ, both s and c evolve from their initial values. The achieved nondimensional value $s = 1$ indicates that the length scale is well-chosen and that it can also represent the distance over which the concentration travels over fresh adsorbent before being completely used up.

The starting condition for our numerical scheme is then taken as the leading order solution

$$Q = \tau - 1 \,, \qquad s = 1 \,, \tag{31}$$

$$c = 1 - Pe^{-1} + Pe^{-1} e^{-Pe(1-x)} - x \,. \tag{32}$$

In terms of the original nondimensional variables,

$$q = t - t_e \,, \qquad s = 1 \,, \tag{33}$$

$$c = 1 - Pe^{-1} + Pe^{-1} e^{-Pe(1-x)} - x \,, \tag{34}$$

for $t > t_e$.

5 Numerical Solution

For most of the processes, the problem is a free boundary problem, and adsorption only occurs in the growing region $x \in [0, s(t)]$ or, in other words, in the region where $c(x, t)$ is strictly larger than 0. In the case of one-dimensional Stefan problems, where the space domain of the problem also evolves with time, a popular approach to deal with the moving boundary is to map the problem to a fixed domain by means of the Landau transformation [21–23]. Other methods to deal with moving boundaries are the phase-field approach and moving mesh methods (see, for instance, [24, 25]). We take a different approach here by rewriting the problem as follows:

$$\delta \frac{\partial c}{\partial t} + \frac{\partial c}{\partial x} = Pe^{-1} \frac{\partial^2 c}{\partial x^2} - \frac{\partial q}{\partial t}, \tag{35}$$

$$\frac{\partial q}{\partial t} = (1 - q) H(c), \tag{36}$$

where $H(c)$ represents the unit step function

$$H(c) = \begin{cases} 1 & \text{for} \quad c > 0, \\ 0 & \text{otherwise}. \end{cases} \tag{37}$$

The function $H(c)$ ensures that $q(x, t)$ only increases in the regions where CO_2 is present. Because we have effectively removed the moving boundary $x = s(t)$, we can modify the boundary condition at $x = s(t)$ in (11) to read

$$\left. \frac{\partial c}{\partial x} \right|_{x=l} = 0, \tag{38}$$

where $l = L/\mathcal{L}$. We then use standard central finite differences in space and explicit Euler in time. The position of the free boundary $s(t)$ can be simply estimated by checking which is the first point of the mesh that satisfies $c(x, t) = 0$ at a given time step.

6 Results and Discussion

In this section, we present a set of results for the concentration and adsorption of CO_2 passing through a column packed with the adsorbent material. We take the appropriate experimental conditions from the work of [15], as shown in Table 1.

In Fig. 2, we plot the CO_2 concentration and adsorption profiles along the column at six different times. For small times, the concentration and the adsorption have approximately linear and constant profiles, respectively (consistent with the small time asymptotic expressions (33)–(34)). After the early time transient, both concentration and adsorption develop a front, with almost exactly the same shape, which travels in a self-similar form along the column. This behaviour was predicted via the travelling wave solution (20)–(22) (circles in Fig. 2), which shows excellent agreement with the numerical curves.

Note that the small discrepancy for small values of x in the curves corresponding to $t = 2$ is expected since the travelling wave solution is not valid for small times. So, the discrepancy simply indicates that the wave has not yet fully developed a steady shape. Around $t \approx 8.75$, the gas front reaches the end of the column for the

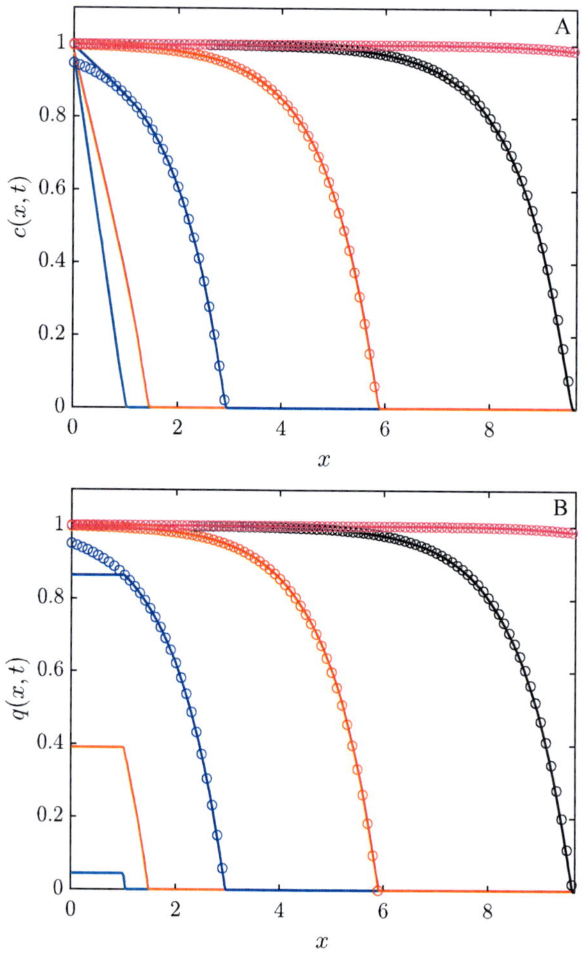

Fig. 2 Concentration (**a**) and adsorption (**b**) along the x-axes of the column bed from the numerical (solid lines) and travelling wave (circles) solutions at $t = 0.05, 0.5, 2, 5, 8.75, 12.92$ (from left to right)

first time, indicating that the process has almost finished. The final curve, for time $t = 12.92$, shows that virtually all the CO_2 is escaping.

In Fig. 3a, the concentration curves at the inlet, centre and end of the column are shown. The numerical solution at the inlet begins at $c \approx 0.96$ (since we are using (34) evaluated at $x = 0$ to initialise our numerical scheme) and reaches the inlet value $c = 1$ around $t = 2$. The travelling wave is not expected to function for small times, and we see that it does not capture the inlet curve until around $t = 4$. However, at the middle and end of the column, the travelling wave coincides with the numerical solution.

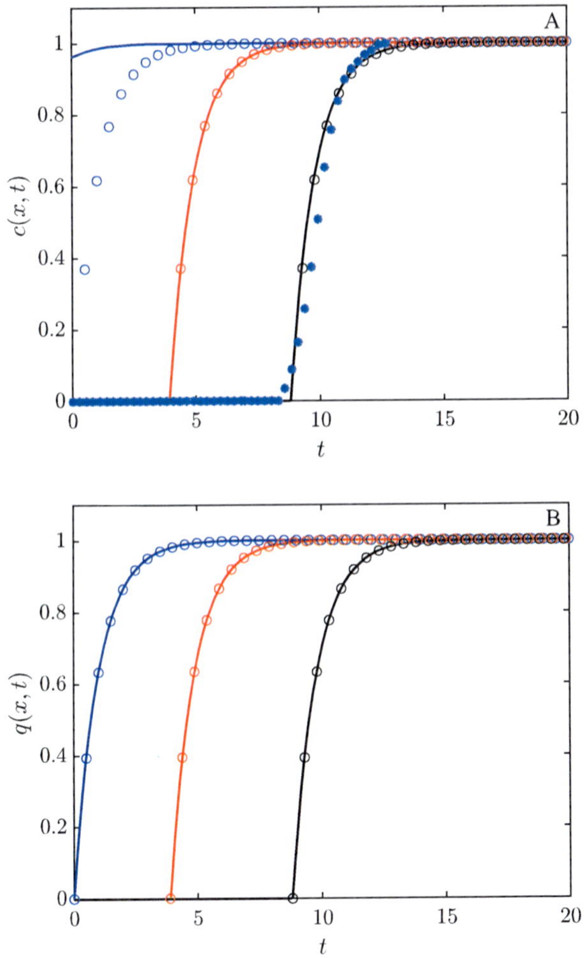

Fig. 3 Time evolution of the concentration (**a**) and the adsorption (**b**) at the column inlet, middle and outlet (from left to right). The solid lines correspond to the numerical solution and the circles to the travelling wave solution (20) and (22). The asterisks represent experimental data points from [15]

A typical experimental measurement is the time evolution of the gas concentration at the outlet, known as the breakthrough curve. This corresponds to the outlet curve in Fig. 3a. Breakthrough starts around $t = 9$, corresponding to a dimensional value of $9\,\tau = 10.7\,\text{min}$, and after a further interval of $\Delta t \approx 5$ (5.95 min), the concentration reaches 90% of the inlet value. The stars represent the experimental data of [15]. The present solutions agree well with this data, although there is some discrepancy near the point of initial breakthrough. This seems to be an issue with the model (rather than the approximate solution method), which will always show the largest gradient at initial breakthrough: the data shows an initial slow increase. However, the fact that the solutions of our model in general show close agreement with experiments makes our model and, in particular, the travelling wave solution a potential tool to understand and so improve practical carbon capture.

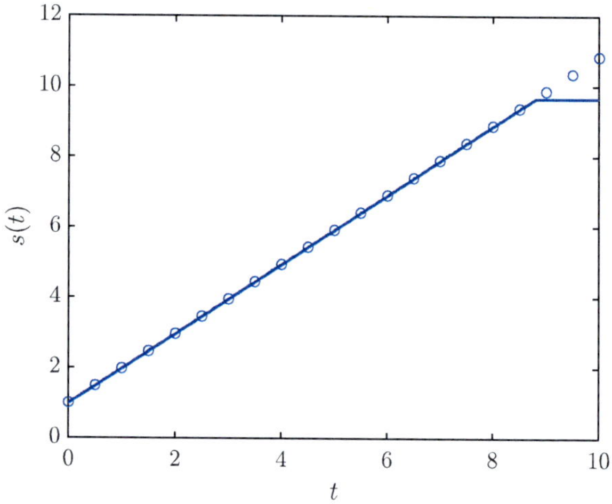

Fig. 4 Position of the free boundary $s(t)$ predicted by the numerical (solid line) and travelling wave (circles) solutions

Figure 3b shows the adsorption curves as a function of time at the inlet, middle and outlet of the column. As predicted, the q curve is similar to that of c, and the travelling wave and numerics agree well for all the three times.

Finally, the position of the free boundary as a function of time is presented in Fig. 4. The agreement between the numerical and the analytical solution (21) is excellent, confirming the velocity estimate $v = 1/(1+\delta)$. The start of the flat region in the numerical solution corresponds to the time when the CO_2 reaches the column outlet for the first time. The travelling wave solution does not take the end of the pipe into account, so s continues to grow (obviously, we could just stop it at $s = L$).

7 Conclusions

In the present work, we have presented and analysed a mathematical model describing a gas passing through a column filled with an adsorbent porous material. The model consists in governing equations for the averaged concentration and averaged adsorbate in a cylindrical geometry, as derived in [18]. A travelling wave solution was found and compared to the numerical solution, showing excellent agreement. Our model solutions show good agreement with data from a carbon capture experiment using a CO_2–N_2 gas mixture, indicating that the travelling wave solution can be an excellent tool to help understand the carbon capture process and hopefully aid in the design of future equipment.

Acknowledgments Francesc Font acknowledges financial support from the *Juan de la Cierva* programme (grant IJC2018-038463-I) from the Spanish MICINN and from the Obra Social 'la Caixa' through the programme Recerca en Matemàtica Col·laborativa. The authors have been partially funded by the CERCA Programme of the Generalitat de Catalunya. Tim G. Myers acknowledges the support of Ministerio de Ciencia e Innovacion Grant No. MTM2017-82317-P. This project has received funding from the European Union's Horizon 2020 research and innovation programme under grant agreement no. 707658. The authors thank Gloria Garcia for her contribution in the art work of the manuscript.

References

1. Climate change: how do we know? https://climate.nasa.gov/evidence/. Accessed 10 Dec 2019
2. Suarez, C.A., Edmonds, M., Jones, A.P.: Earth catastrophes and their impact on the carbon cycle. Elements **15**(5), 301–306, 10 (2019)
3. Svensen, H., Planke, S., Polozov, A.G., Schmidbauer, N., Corfu, F., Podladchikov, Y.Y., Jamtveit, B.: Siberian gas venting and the end-Permian environmental crisis. Earth Planet. Sci. Lett. **277**(3), 490–500 (2009)
4. Steeneveldt, R., Berger, B., Torp, T.A.: CO_2 capture and storage: closing the knowing–doing gap. Chem. Eng. Res. Des. **84**(9), 739–763 (2006)
5. Figueroa, J.D., Fout, T., Plasynski, S., McIlvried, H., Srivastava, R.D.: Advances in CO_2 capture technology—the U.S. department of energy's carbon sequestration program. Int. J. Greenhouse Gas Control **2**(1), 9–20 (2008)
6. Kanniche, M., Bouallou, C.: CO_2 capture study in advanced integrated gasification combined cycle. Appl. Thermal Eng. **27**(16), 2693–2702 (2007)
7. Thiruvenkatachari, R., Su, S., An, H., Yu, X.X.: Post combustion CO_2 capture by carbon fibre monolithic adsorbents. Prog. Energy Combust. Sci. **35**(5), 438–455 (2009)
8. Ben-Mansour, R., Habib, M.A., Bamidele, O.E., Basha, M., Qasem, N.A.A., Peedikakkal, A., Laoui, T., Ali, M.: Carbon capture by physical adsorption: Materials, experimental investigations and numerical modeling and simulations—a review. Appl. Energy **161**, 225–255 (2016)
9. Dantas, T.L.P., Luna, F.M.T., Silva, I.J., de Azevedo, D.C.S., Grande, C.A., Rodrigues, A.E., Moreira, R.F.P.M.: Carbon dioxide–nitrogen separation through adsorption on activated carbon in a fixed bed. Chem. Eng. J. **169**(1), 11–19 (2011)
10. Li, S., Deng, S., Zhao, L., Zhao, R., Lin, M., Du, Y., Lian, Y.: Mathematical modeling and numerical investigation of carbon capture by adsorption: literature review and case study. Appl. Energy **221**, 437–449 (2018)
11. Rezaei, F., Grahn, M.: Thermal management of structured adsorbents in CO_2 capture processes. Ind. Eng. Chem. Res. **51**(10), 4025–4034 (2012)
12. Shafeeyan, M.S., Daud, W.M.A.W., Shamiri, A.: A review of mathematical modeling of fixed-bed columns for carbon dioxide adsorption. Chem. Eng. Res. Des. **92**(5), 961–988 (2014)
13. Al-Janabi, N., Vakili, R., Kalumpasut, P., Gorgojo, P., Siperstein, F.R., Fan, X., McCloskey, P.: Velocity variation effect in fixed bed columns: a case study of CO_2 capture using porous solid adsorbents. AIChE J. **64**(6), 2189–2197 (2018)
14. Raghavan, N.S., Hassan, M.M., Ruthven, D.M.: Numerical simulation of a PSA system. Part I: isothermal trace component system with linear equilibrium and finite mass transfer resistance. AIChE J. **31**(3), 385–392 (1985)
15. Shafeeyan, M.S., Daud, W.M.A.W., Shamiri, A., Aghamohammadi, N.: Modeling of carbon dioxide adsorption onto ammonia-modified activated carbon: Kinetic analysis and breakthrough behavior. Energy Fuels **29**(10), 6565–6577 (2015)
16. Shen, C., Grande, C.A., Li, P., Yu, J., Rodrigues, A.E.: Adsorption equilibria and kinetics of CO_2 and N_2 on activated carbon beads. Chem. Eng. J. **160**(2), 398–407 (2010)
17. Zhao, Y., Shen, Y., Bai, L., Ni, S.: Carbon dioxide adsorption on polyacrylamide-impregnated silica gel and breakthrough modeling. Appl. Surf. Sci. **261**, 708–716 (2012)
18. Myers, T.G., Font, F., Hennessy, M.G.: Mathematical modelling of carbon capture in a packed column by adsorption. Appl. Energy **278**, 115565 (2020)

19. Bahadur, J., Melnichenko, Y.B., He, L., Contescu, C.I., Gallego, N.C., Carmichael, J.R.: SANS investigations of CO_2 adsorption in microporous carbon. Carbon **95**, 535–544 (2015)
20. Bird, R.B., Stewart, W.E., Lightfoot, E.N.: Transport Phenomena, 2nd edn. Wiley, London (2007)
21. Font, F., Afkhami, S., Kondic, L.: Substrate melting during laser heating of nanoscale metal films. Int. J. Heat Mass Transf. **113**, 237–245 (2017)
22. Font, F.: A one-phase Stefan problem with size-dependent thermal conductivity. Appl. Math. Model. **63**, 172–178 (2018)
23. Font, F., Bresme, F.: Transient melting at the nanoscale: a continuum heat transfer and nonequilibrium molecular dynamics approach. J. Phys. Chem. C **122**(30), 17481–17489 (2018)
24. Boettinger, W.J., Warren, J.A., Beckermann, C., Karma, A.: Phase-field simulation of solidification. Ann. Rev. Mater. Res. **32**(1), 163–194 (2002)
25. Segal, G., Vuik, K., Vermolen, F.: A conserving discretization for the free boundary in a two-dimensional Stefan problem. J. Comput. Phys. **141**(1), 1–21 (1998)

Diffusion Processes at Nanoscale

Claudia Fanelli

Abstract It is well-known that many properties of nanoparticles, such as luminescence, photostability, optical radiation efficiencies and electric properties among others, are size dependent. Hence, the ability to create nanoparticles of a specific size is crucial. Starting from a mathematical description of the nanoparticle growth process and guidelines for efficient growth strategies, I will show a specific practical application of nanoparticles, namely targeted drug delivery. The physical situation modelled involves the motion of a non-Newtonian nanofluid subject to an external magnetic field and an advection–diffusion equation for the concentration of the nanoparticles in the fluid. The ultimate goal is to determine strategies to maximise drug delivery to a specific site.

1 Why Nanoscale Is so Interesting?

When we refer to the nanoscale, we are talking about working with units of matter with dimensions between 1 and 100 nanometres (nm) which are called nanoparticles (NPs). During the last decades, it has been shown that nanotechnology can develop materials and devices that will change the way to work and manage scientific challenges. Nanoparticles have unique properties that naturally occur at that scale. There are two main features that show the power of nanoscale:

1. **Surface area-to-volume ratio**: materials made up of nanoparticles have a greater surface area when compared to the same volume of material made up of larger particles. This means that a great amount of the material can come into contact with surrounding materials, increasing the reactivity.
2. **'Tunability' of properties**: with slight changes in the size of nanoparticles, a scientist is able to control and adapt a material property.

C. Fanelli (✉)
Universitat Politècnica de Catalunya, Barcelona, Spain

F. Font, T. G. Myers (eds.), *Multidisciplinary Mathematical Modelling*,
SEMA SIMAI Springer Series 11, https://doi.org/10.1007/978-3-030-64272-3_4

A very famous example that shows the great potential of this scale is the case of gold nanoparticles. It is said that *all that glitters is not gold*, but what I will ask you is: all the gold glitters? At the nanoscale, the motion of the gold's electrons is confined and, because of that, they react differently with light compared to larger scales. The result is that gold nanoparticles are not yellow as we expect but can appear in red or purple according to their size. Moreover, adjusting their dimensions, gold nanoparticles can be tuned according to the purpose; for example, they can selectively accumulate in tumours in order to identify diseased cells and target laser destruction of the tumour avoiding healthy cells.

In Sect. 2, we will start explaining how to deal with the manufacturing of nanoparticles, optimising the different processes involved in NP growth in solution through a mathematical description of the growth process and guidelines for efficient growth strategies. Finally, in Sect. 3, we will show preliminary results on a specific practical application of NPs, namely targeted drug delivery.

2 Nanoparticles Growth

Many properties at nanoscale, such as electronic and optical properties of metals and semiconductors, luminescence and photostability and optical radiation efficiencies among others, are size dependent [13]. It is clear by now that the ability to create nanoparticles of a specific size is crucial. In order to do this, we need a clear understanding for the process of growing nanoparticles.

Nanoparticles can be prepared by both gas phase and solution-based synthesis techniques. Although the first method can produce large quantities of nanoparticles, agglomeration and nonuniformity in particle size and shape are typical problems [7]. Using the precipitation method (i.e. the creation of a solid from a solution) monodisperse spherical nanoparticles can be generated. The standard approach is to apply the classical La Mer and Dinegar synthesis strategy where nucleation and growth are separated [6]. The strategy is to rapidly add the precursor at high temperature to batch reactors, causing a short nucleation burst in order to create a large number of nuclei in a short space of time. The seeds generated are used for the latter particle growth stage, which involves two different stages:

1. The focusing period, where the particles increase rapidly and the size distribution is relatively small.
2. The defocusing period, where the growth slows down and the size distribution becomes larger.

The first phase leads to the desired result of monodisperse nanoparticles. In the second phase, we can observe a phenomenon called *Ostwald ripening* (OR), a process by which larger particles grow at the expense of the smaller ones that dissolve due to their much higher solubility.

This process, represented in Fig. 1, produces monomer, which is subsequently used to support the growth of the larger particles. However, this simultaneous growth

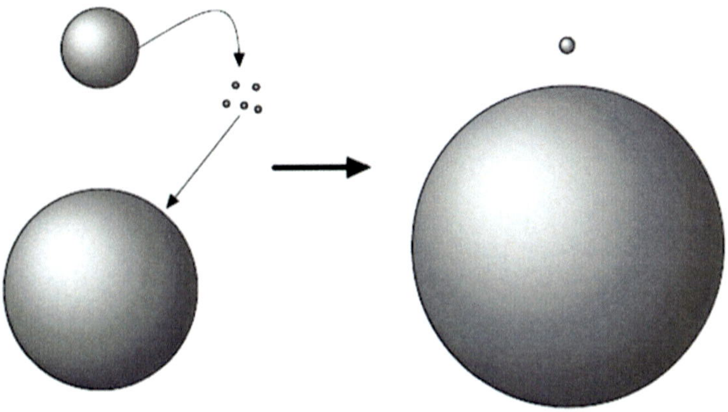

Fig. 1 Sketch of Ostwald ripening in the case of two particles with different radius

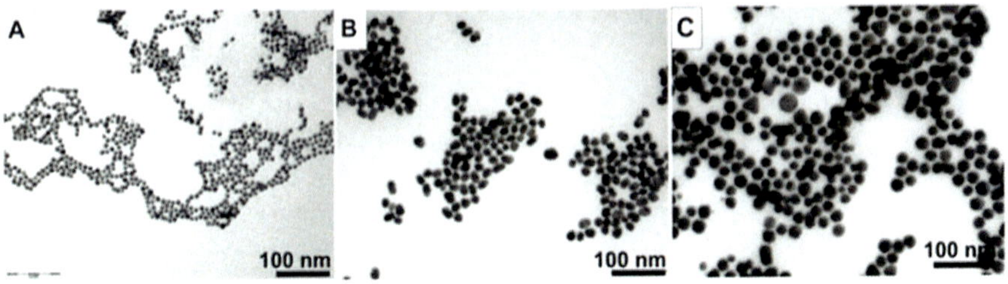

Fig. 2 Microscopic images of the growth of gold seed particles at different time steps from the supplementary material of Bastús et al. [2]

and dissolution leads to the unwanted defocusing of the particle size distribution (PSD). Recently, it has been shown that the PSD can be refocused by changing the reaction kinetics. For example, in Fig. 2, we can see three snapshots of the growth process of gold nanoparticles from Bastús et al. [2], where temperature, gold precursor to seed particle concentration, and pH are adjusted during the process in order to obtain the desired result.

In order to be able to model this process, we will assume that the system is dilute, such that particle interaction and aggregation are neglected. Moreover, the growth is always spherically symmetric, and the effect of any solvent used to facilitate the growth process is accounted for by the diffusion and or kinetic rate constant.

2.1 Model for a Single Particle

From a mathematical point of view, the growth of single nanoparticle is analogous to a one-phase Stefan problem.

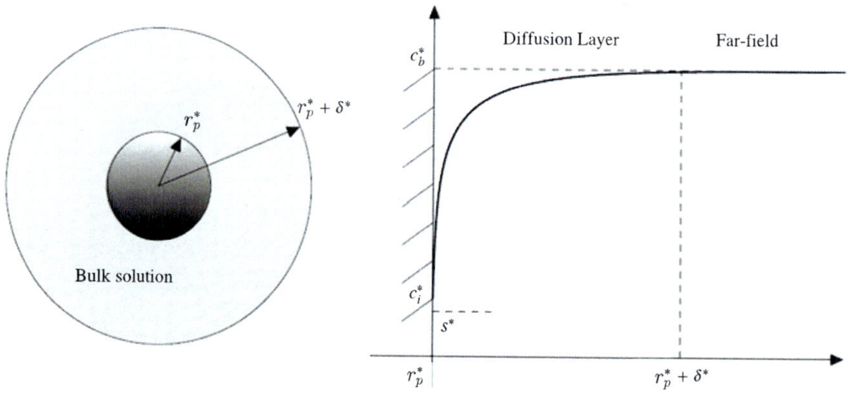

Fig. 3 Schematic of a single nanoparticle with radius r_p and the surrounding monomer concentration profile where s, c_i and c_b are the particle solubility, the concentration at the surface of the particle and the far-field concentration, respectively

As represented in Fig. 3, the growth occurs thanks to the diffusion of the monomer molecules from the bulk to the surface of the nanoparticles of radius r_p. Monomers are brought to the particle surface via diffusion from the bulk solution at a rate determined by a diffusion coefficient D. The monomer concentration $c(r, t)$ follows the diffusion equation described by

$$\frac{\partial c}{\partial t} = \frac{D}{r^2}\frac{\partial}{\partial r}\left(r^2 \frac{\partial c}{\partial r}\right), \qquad r_p < r < \delta, \tag{1}$$

where D is the diffusion coefficient and r is the distance from the centre of the particle with radius r_p. To conform to the standard literature, we have included a diffusion layer of width δ around the particle, where the concentration adjusts from c_i, which is the monomer concentration adjacent to the surface, to c_b, the far-field concentration. We can write the boundary condition at the particle surface in terms of the solubility as

$$c(r, t) = c_i = s + \frac{D}{k}\frac{\partial c}{\partial r}\bigg|_{r=r_p}, \tag{2}$$

where s is given by the Ostwald–Freundlich equation

$$s = s_\infty \exp\left(\frac{2\sigma V_M}{r_p R_g T}\right) = s_\infty \exp\left(\frac{\alpha}{r_p}\right), \tag{3}$$

where s_∞ is the solubility of the bulk material, V_M is the molar volume of monomer in solution, σ is the interface energy, R_g is the universal gas constant and T is the absolute temperature. The parameter $\alpha = 2\sigma V_M/(R_g T)$ is called the capillary length, and it is usually of the order of 1–6 nm. The variation of the particle solubility is generally neglected based on the assumption that $\alpha \ll r_p$, but this approximation is invalid for nanoparticles where the capillary length is of the same order of magnitude as the particle radius. In fact, despite its minor contribution to the growth of a single particle it does play an important role in the growth of a group of nanocrystals.

At the far field, we can approximate the boundary condition via a mass conservation relation as

$$c(r+\delta) = c_b(t) \approx c_0 - \frac{4\pi N_0}{3V_M} {r_p}^3, \tag{4}$$

where N_0 is the population density.

The initial value of the concentration is equal to c_0 and represents the initial well-mixed and uniform concentration of the monomer solution.

The particle radius r_p is also an unknown function of time that can be determined by balancing mass added to the crystal with the incoming monomer flux. Thus, we compare the flux in terms of the change in volume of a particle obtaining the Stefan condition

$$\frac{\mathrm{d}r_p}{\mathrm{d}t} = V_M D \frac{\partial c}{\partial r}\bigg|_{r=r_p}, \qquad r_p(0) = r_{p,0}, \tag{5}$$

where $r_{p,0}$ is the initial particle radius.

In order to obtain a good approximation of the solution of this model, we need to make some assumptions. The equation that describes the evolution of the concentration over time contains the unknown δ, which is more commonly termed as the boundary layer. Diffusion boundary layers are time dependent and, if the fluid is initially well-mixed, $\delta(0) = 0 < r_p$. Therefore, the standard model is not valid for early times, and we have to consider the approximation $\delta \gg r_p$. Diffusion occurs over a much faster time-scale than growth, that is, the diffusion time-scale is much smaller than the growth time-scale $t_D \ll t_G$. Hence, the standard pseudo-steady approximation will be accurate.

As explained in detail in Myers and Fanelli [9], it can be shown that the explicit equation

$$r_p = \frac{r_m}{2} \frac{\left[1 + 2f(r_{p0})\exp\left(\frac{t-t_0}{G}\right) - \sqrt{-3 + 12 f(r_{p0})\exp\left(\frac{t-t_0}{G}\right)}\right]}{\left[-1 + f(r_{p0})\exp\left(\frac{t-t_0}{G}\right)\right]} \tag{6}$$

is a good approximation of the behaviour of the nanoparticle radius. It is interesting to notice from the parameter $G = (ak + bD)/(6ab(akbD))$ that ak and bD are interchangeable: it does not matter if we define them the opposite way round, the result is the same. Physically this means that the model cannot distinguish between diffusion and reaction driven growth. Moreover, this solution only depends on two independent parameters, G and r_m, strongly reducing the errors in the fitting process.

2.2 Ostwald Ripening

Although considering the solution given by Eq. (6) as the behaviour of the mean radius of a system of nanoparticles will give us an excellent agreement with experimental data, in order to control the undesired Ostwald ripening, we need to keep track of the radius of each particle.

Therefore, we now consider a system of N particles that follow a normal standard distribution with mean initial radius $\bar{r}_0$ and initial standard deviation σ_0. Since there are no interparticle diffusional interactions, we consider the same equations obtained before for each particle in the system, paying attention to the equation for the bulk concentration c_b. In fact, now the mass balance used in order to obtain the value of the concentration at the far field will be approximate as

$$c_b(t) \approx c_0 - \frac{4\pi N_0}{3NV_M} \sum_{i=1}^{N} r_i^3. \tag{7}$$

where we called r_i the ith particle radius.

For certain materials, the Ostwald ripening may take a very long time and so be difficult to observe. To demonstrate that the current model can predict OR, we investigate the simplest possible case. In Fig. 4, the evolution of the radius and the solubility of two particles is shown, using the parameters from Fanelli et al. [3]. However, since we are using data for CdSe, we anticipate a slow process.

The system is defined choosing the initial radii to be 2 and 2.5 nm. The two governing equations may be easily solved using the MATLAB ODE solver `ode15s`. The first figure shows the evolution of the radii for more than 25 h. The solid line represents the evolution of the 2.5-nm particle, the dashed line is the 2-nm one. As can be seen, for small times, both particles grow rapidly, and however,

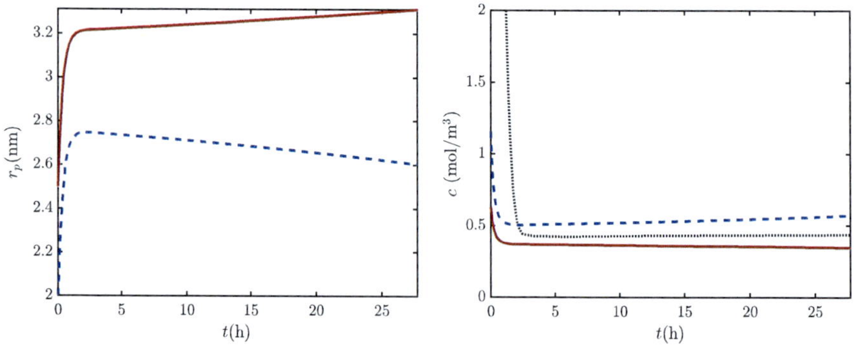

Fig. 4 Evolution of two CdSe nanoparticles: (left) Change in time of the radii of two particles with initial radii of 2 nm (dashed line) and 2.5 nm (solid line). (right) Change in bulk concentration (dotted line) and solubility of smaller (dashed line) and larger (solid line) particles

after around 1.7 h, the smaller particle starts to shrink, while the larger one grows linearly. In the second figure, the variation of the particle solubility and the bulk concentration is shown, where solid and dashed lines correspond to the 2.5- and 2-nm particles' solubility, respectively, while the dotted line is the bulk concentration. With reference to the variation of the radius, it is clear that the rapid growth phase corresponds to a sharp decrease in the bulk concentration. Initially the solubility of each particle is below the bulk concentration and decreases as r_p increases. The Ostwald ripening begins when the solubility of the smaller particle crosses the c_b curve, at $t \simeq 1.7$h, and subsequently its size decreases. The solubility of the larger particle keeps slowly decreasing and remains below the bulk concentration until the end of the simulation.

2.3 N Particles System

To compare the solution of our model with the experimental data from Peng et al. [12], we consider a distribution of N nanoparticles (with $N = 10$ and $N = 1000$), where the initial distribution is generated by random numbers, with an initial mean radius $\bar{r}_{i,0}$ of 2.92 nm and a standard deviation of $\sigma_o = 8.9\%$. In the numerical solution, if a particle decreases below 2 nm, it is assumed to break up, and all the monomers return to the bulk concentration.

In practice, N would be much higher. The population density is given as $N_0 = 8.04 \times 10^{21}$ crystals/m^3, so in a volume $V \approx 7 \times 10^{-6}$m^3, we would expect around 10^{16} crystals. In the experimental data used here, extra monomer is added to the solution after 3 h, and we stop our simulation then. In Fig. 5, it is clear that the approximation given by the model that takes into account all the particles in the system is very accurate, since it approaches the experimental data represented by

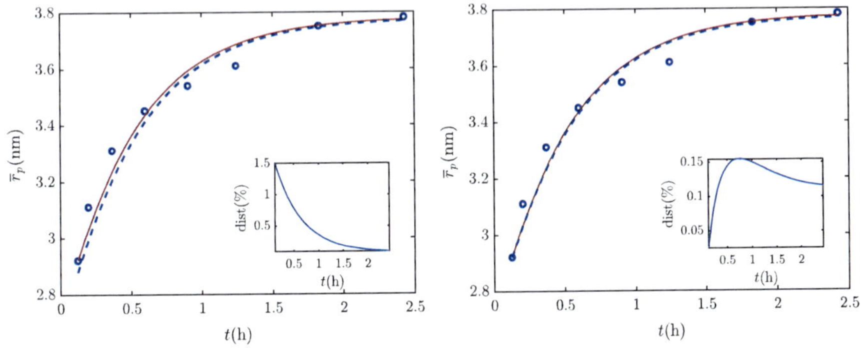

Fig. 5 Comparison of the model for N particles (dashed lines) from Fanelli et al. [3] with experimental data from Peng et al. [12] (dots) using $N = 10$ in (**a**) and $N = 1000$ in (**b**). The solid lines represent the explicit solution for the one particle model, Eq. (6). The inset plots show the percentage difference between the models

the black dots during the whole process. Moreover, it is able to track the Ostwald ripening effect in order to adjust kinetic reactions and preserve the monodispersity result.

3 Drug Delivery

In order to show a practical use of the ability to control the size of nanoparticles, it is interesting to understand a very popular and effective way to operate the injection of drugs in human body. One of the biggest motivations, taking the case of cancer therapy, is that the main approaches are non-specific and their efficacy is reduced. Since normally very high doses of drugs are required to target malignant cells, in order to specifically target tumours, there are currently two standard techniques. The first involves the inhibition, by various means, of drug delivery to healthy non-cancerous cells, while the second involves the direct conduction of drugs into the tumour site. It is now accepted that one way to achieve the second technique is by using nanoparticles to deliver the drugs directly to the tumour cells, which results in minimum drug leakage into normal cells [1, 14].

The technique of magnetically targeted drug delivery involves binding a drug to small biocompatible magnetic NPs, injecting them into the bloodstream and then using a high-gradient magnetic field to direct them to the target region. The movement and directing of such particles are the focus of this study.

As represented in Fig. 6, a key issue in the magnetic drug delivery is whether the applied magnetic forces can compete with convective blood (drag) forces that tend to wash particles away. The blood flow drag forces on the particle vary with its position in the blood vessel. A particle at the vessel centreline will experience a higher blood velocity and hence a higher drag force, but a particle near the blood vessel wall will be surrounded by a near-zero blood velocity. This decrease in velocity is due to the flow resistance provided by the vessel wall, the 'no-slip' boundary condition. Thus, a particle near the vessel wall will experience a much smaller drag force and can potentially be held by a much smaller magnetic force.

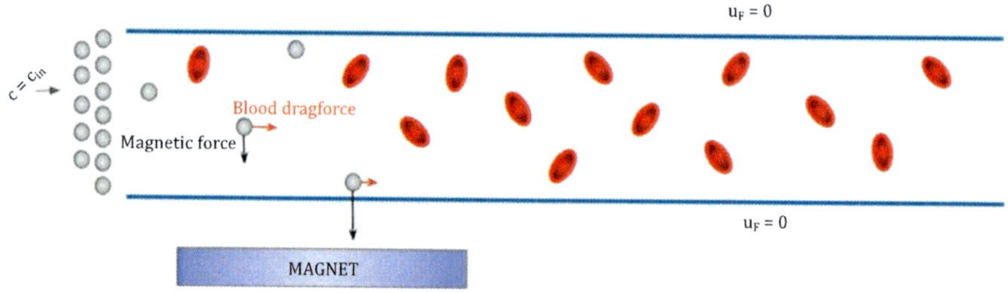

Fig. 6 Schematic of the injection of magnetic nanoparticles in a vessel subjected to a magnetic field

Blood containing nanoparticles may be considered as a form of non-Newtonian nanofluid. A nanofluid is a suspension of nanoparticles in a base fluid, and its motion is usually described by a coupled set of partial differential equations describing the fluid flow and the particle concentration. In the next two sections, we will analyse first the motion of the fluid and then its effect on the nanoparticles concentrated in it.

3.1 Non-Newtonian Fluids

In order to understand the issues of this model, let us start from some general definition. In fluid dynamics, the viscosity of the fluid is a measure of its resistance to deformation by shear stress. We can note that in Newton's law of viscosity,

$$\boldsymbol{\tau} = \eta \nabla \mathbf{u} \,, \tag{8}$$

the shear stress $\boldsymbol{\tau}$ has units equivalent to a momentum flux, i.e. momentum per unit time per unit area. According to Newton's law of viscosity, this momentum flow occurs across a velocity gradient, and the magnitude of the corresponding momentum flux is determined by the viscosity η, which is constant.

We can define a Newtonian behaviour under constant temperature and pressure following the description from Owen and Phillips [11] which states:

- The only stress generated in simple shear flow is the shear stress τ, and the two normal stresses being zero.
- The shear viscosity does not vary with the shear rate.
- The viscosity is constant with respect to the time of shearing, and the stress in the liquid falls to zero immediately when the shearing is stopped.
- The viscosities measured in different types of deformation are always in a simple proportion to one another.

Any liquid that deviates from the above behaviour is said to be non-Newtonian. Typical examples of non-Newtonian fluids in nature are most of the body fluids, such as blood, saliva, goopy eye fluid, etc. We can also define a fluid as non-Newtonian if the stress tensor cannot be expressed as linear function of the components of the velocity gradient.

Therefore, the relation between the shear stress and the shear rate can be written as

$$\boldsymbol{\tau} = \eta(\dot{\gamma})\dot{\gamma} \,, \tag{9}$$

where $\dot{\gamma}$ is the shear rate. In this study, three famous models for a non-Newtonian fluid are compared. They are characterised by the following expressions:

1. **The power-law model**
   The standard power-law model describes the viscosity by

$$\eta_p = K\, |\dot{\gamma}|^{n_p - 1}\,, \tag{10}$$

   where K is constant and $\dot{\gamma}$ is the shear rate. If $n_p < 1$, the fluid is pseudoplastic or shear thinning, if $n_p > 1$, it is dilatant or shear thickening.
2. **The Carreau model**
   The Carreau model describes the viscosity by

$$\eta_c = \eta_\infty + (\eta_0 - \eta_\infty)\left[1 + \lambda^2\dot{\gamma}^2\right]^{(n_c-1)/2}\,, \tag{11}$$

   where λ is a constant and η_0 and η_∞ are the limiting viscosities at low and high shear rates, respectively.
3. **The Ellis model**
   The Ellis model describes the viscosity in terms of the shear stress as

$$\frac{1}{\eta_e} = \frac{1}{\eta_0}\left(1 + \left|\frac{\tau}{\tau_{1/2}}\right|^{\alpha-1}\right)\,, \tag{12}$$

   where η_0 is the viscosity at zero shear and $\tau_{1/2}$ is the shear stress at which the viscosity is $\eta_0/2$. We have to notice that this model cannot predict the second Newtonian plateau typical of the shear-thinning fluids.

The velocity profiles and the corresponding viscosity profiles for blood obtained from the four different models are compared in Fig. 7.

We can observe that, even if the parameters are chosen in order to obtain a good agreement for the velocity behaviours, the relative viscosity solutions can differ drastically. The second figure shows clearly that, for example, the viscosity of the power-law model, represented by the dashed–dotted line, tends to infinity as $y \to 0$, while the rest of viscosities have a very different value there. As we can see from our simulations, both Ellis and Carreau models give similar approximations for the blood flow. The Carreau model is generally preferred due to its ability to predict the shear-thinning behaviour of the blood (see for example [5, 10]). However, its expression for the variation of the blood velocity cannot be integrated and we will have to solve it numerically, while the Ellis model can be treated analytically.

For this reason, the viscosity described from Eq. (12) will be the one used in this study, where Navier–Stokes equations for a fluid in a channel under lubrication assumptions are solved in order to obtain an explicit equation for the motion of the blood in the vessel.

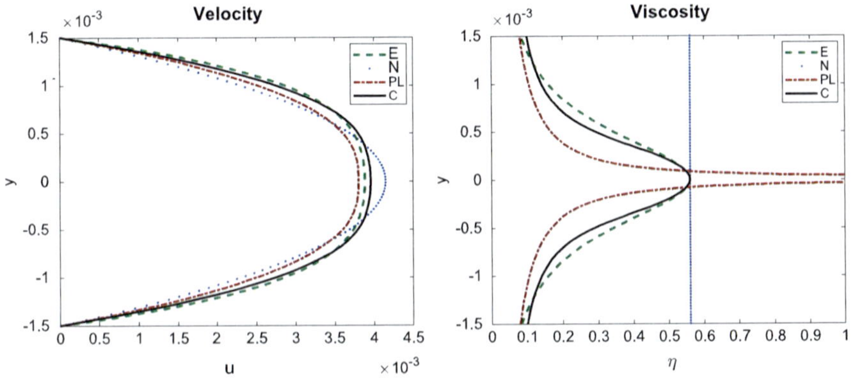

Fig. 7 Comparison of the velocity (left) and viscosity (right) of Newtonian (dotted line), power-law (dashed–dotted line), Carreau (solid line) and Ellis (dashed line) models, where the parameters are taken from [8]

3.2 Advection–Diffusion Equation

The behaviour of the concentration of magnetic nanoparticles in the bloodstream is obtained following the model developed by Grief and Richardson [5]. The model takes into account the Stokes drag and the magnetic forces but also the interactions and collisions between moving red blood cells in the bloodstream which cause a diffusive motion of the particles.

Therefore, the governing equation describing the motion of magnetic particles in the blood stream is an advection–diffusion equation for the particles concentration:

$$\frac{\partial c}{\partial t} + \nabla \cdot \left[(\mathbf{u}_F + \mathbf{u}_p)c\right] = \nabla \cdot (Dc) \ , \tag{13}$$

where $\mathbf{u}_F$ is the solution of the equations resulting from the model described in Sect. 3.1, $\mathbf{u}_p$ is the particle velocity, which is found by balancing hydrodynamic and magnetic forces, and D is the diffusion coefficient. We can use the definition of the Stokes drag, which is the force of viscosity on a spherical particle of radius a moving through a viscous fluid, obtaining

$$\mathbf{u}_p = \frac{\mathbf{F}_{\text{mag}}}{6\pi a\, \eta(\dot{\gamma})} \ , \tag{14}$$

where $\mathbf{F}_{\text{mag}}$ is the magnetic force and $\eta(\dot{\gamma})$ is the blood shear rate. We will consider the diffusion due to the Brownian motion negligible (cf. [10], $D_{\text{Br}} \approx 6 \times 10^{-13}$), and then the diffusion coefficient D is given only by the contribution of the shear-induced diffusion

$$\mathbf{J}_{\text{diff}} = -D\nabla c = -\left(K_{\text{sh}}\,(r_{\text{RBC}})^2\,\dot{\gamma}\right)\nabla c\,, \tag{15}$$

where K_{sh} is a dimensionless coefficient that depends on the blood cell concentrations and r_{RBC} is the blood cell radius. The shear-induced diffusion is due to the fact that the red blood cells suspended in plasma collide with each other causing random motion with a diffusive character.

At first, we will consider that particles can flow out through the walls with a certain permeability coefficient κ, which implies to set Robin boundary conditions at the top and the bottom of the channel,

$$\left(u_p c - D\frac{\partial c}{\partial y}\right)\bigg|_{y=\pm R} = \kappa c. \tag{16}$$

We also need to specify the inlet and the outlet conditions, assuming that the flux entering in the channel is constant and equal to the inlet concentration c_{in}, while at the end of the channel, we are sufficiently far from the magnet, hence,

$$c(x, y, t)|_{x=0} = c_{\text{in}}, \qquad \frac{\partial c}{\partial x}\bigg|_{x=L} = 0\,. \tag{17}$$

The transient problem is finally well-posed with the initial condition $c(x, y, 0) = 0$.

3.3 Preliminary Results

It has been shown in [5] that for sufficiently weak fields on superparamagnetic particles (i.e. with a diameter smaller than 30 nm), we can consider the approximation

$$\mathbf{F}_{\text{mag}} \approx \frac{m_{\text{sat}}^2}{6kT}|\mathbf{B}|^2\,, \tag{18}$$

where m_{sat} is the saturation magnetisation of the magnetic particle, k is the Boltzmann's constant, T is the absolute temperature and $\mathbf{B}$ is the magnetic field.

Figure 8 shows the preliminary results of the simulation of the movement of magnetic nanoparticles in the bloodstream in non-dimensional variables. The velocity vector field and the concentration of magnetic nanoparticles entering in the vessel are shown. In the plots, on the top, we have that $F_{\text{mag}} = 0$, and we can observe how the nanoparticles are flushed away with the fluid and transported through the vessel just due to the convective flow of the blood. On the other hand, the plots on the bottom show the case where a weak magnetic field is applied, and its effect on the particles is clearly visible. As shown in Fig. 6, the magnet is located at

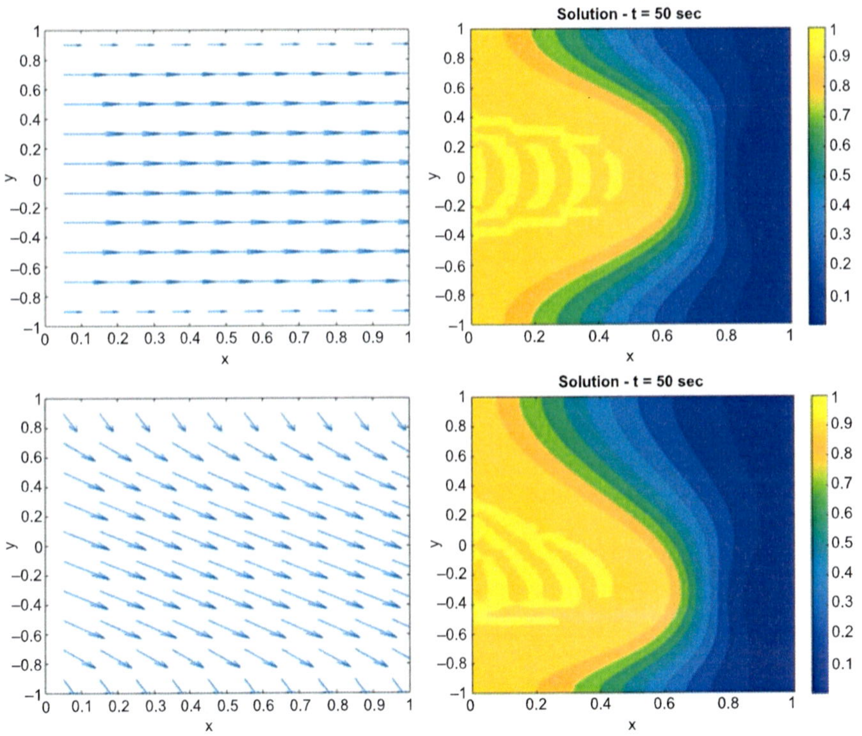

Fig. 8 The velocity vector field (left) and the concentration of magnetic nanoparticles entering in the vessel (right) are shown. Figures on the top correspond to $F_{\mathrm{mag}} = 0$, while figures on the bottom show the case in which a weak magnetic field is applied

the bottom of the vessel and is able to attract them, changing their directions towards the lower border of the domain. Further details can be found in the recent work of Fanelli et al. [4]. However, open challenges on the optimal location and strength of the magnetic field are still unsolved and are meant to be studied in the future work.

4 Conclusions

The goal of this study was to present two innovative processes in the developing field of nanotechnology.

In the first part, we analysed the growth of nanoparticles in solution by means of mathematical tools. Firstly, we studied the simplified problem of the growth of a single particle, deriving the correct assumptions for the standard model and obtaining a new analytical solution in order to approximate the evolution for the radius of the particle. Secondly, we extended the model for a system of N particles, where N is arbitrarily large, in order to control the undesired effects of Ostwald ripening. Excellent agreement with experimental data was obtained for both models.

In the second part, we focused on a practical application in nanomedicine named magnetic drug targeting. The importance to consider the shear-thinning behaviour of the blood when modelling this medical application was demonstrated. Once obtained an analytical solution for the flow of the blood thanks to simplifications in the geometry, the transport and diffusion of the drugs bound to magnetic nanoparticles into the bloodstream was analysed. Moreover, two preliminary results of simulations obtained applying a constant magnetic field were shown.

In both cases, it is shown how a mathematical approach can be a powerful tool in order to optimise the use of nanoparticles for the specific industrial application.

Acknowledgments I would like to thank Dr. Katerina Kaouri and Prof. Tim Phillips from Cardiff University which hosted me for three months, giving me the chance to work with them on the drug delivery problem. I would also like to thank my PhD supervisor Prof. Tim Myers for collaborating in the whole study.

References

1. Bahrami, B., Hojjat-Farsangi, M., Mohammadi, H., Anvari, E., Ghalamfarsa, G., Yousefi, M., Jadidi-Niaragh, F.: Nanoparticles and targeted drug delivery in cancer therapy. Immunol. Lett. **190**, 64–83 (2017)
2. Bastús, N.G., Comenge, J., Puntes, V.: Kinetically controlled seeded growth synthesis of citrate-stabilized gold nanoparticles of up to 200 nm: size focusing versus Ostwald ripening. Langmuir **27**(17), 11098–11105 (2011)
3. Fanelli, C., Cregan, V., Font, F., Myers, T.G.: Modelling nanocrystal growth via the precipitation method. Int. J. Heat Mass Transf. **165**, 120643 (2021)
4. Fanelli, C., Kaouri, K., Phillips, T.N., Myers, T.G., Font, F.: Magnetic nanodrug delivery in non-Newtonian blood flows. Submitted. arXiv:2102.03911
5. Grief, A.D., Richardson, G.: Mathematical modelling of magnetically targeted drug delivery. J. Magn. Magn. Mater. **293**(1), 455–463 (2005)
6. La Mer, V.K., Dinegar, R.: Theory, production and mechanism of formation of monodispersed hydrosols. J. Am. Chem. Soc. **72**(11), 4847–4854 (1950)
7. Mantzaris, N.V.: Liquid-phase synthesis of nanoparticles: particle size distribution dynamics and control. Chem. Eng. Sci. **60**(17), 4749–4770 (2005)
8. Myers T.G.: Application of non-Newtonian models to thin film flow. Phys. Rev. E **72**(6), 066302 (2005)
9. Myers T.G., Fanelli C.: On the incorrect use and interpretation of the model for colloidal, spherical crystal growth. J. Colloid Interface Sci. **536**, 98–104 (2019)
10. Nacev A., Beni C., Bruno O., Shapiro B.: The behaviors of ferromagnetic nano-particles in and around blood vessels under applied magnetic fields. J. Magn. Magn. Mater. **323**(6), 651–668 (2011)
11. Owens, R.G., Phillips, T.N.: Computational Rheology. World Scientific, Singapore (2002)
12. Peng, X., Wickham, J., Alivisatos, A.P.: Kinetics of II–VI and III–V colloidal semiconductor nanocrystal growth: "focusing" of size distributions. J. Am. Chem. Soc. **120**(21), 5343–5344 (1998)

13. Viswanatha, R., Sarma, D. Growth of Nanocrystals in Solution. Nanomaterials Chemistry: Recent Developments and New Directions, pp. 139–170. Wiley, London (2007)
14. Wong, K.V., De Leon, O. Applications of nanofluids: current and future. Adv. Mech. Eng. **2**, 519659 (2010)

Maximum Likelihood Estimation of Power-Law Exponents for Testing Universality in Complex Systems

Víctor Navas-Portella, Álvaro González, Isabel Serra, Eduard Vives, and Álvaro Corral

Abstract Power-law-type distributions are extensively found when studying the behavior of many complex systems. However, due to limitations in data acquisition, empirical datasets often only cover a narrow range of observations, making it difficult to establish power-law behavior unambiguously. In this work, we present a statistical procedure to merge different datasets, with two different aims. First, we obtain a broader fitting range for the statistics of different experiments or observations of the same system. Second, we establish whether two or more different systems may belong to the same universality class. By means of maximum

V. Navas-Portella (✉)
Centre de Recerca Matemàtica, Bellaterra, Barcelona, Spain

Barcelona Graduate School of Mathematics, Barcelona, Spain

Facultat de Matemàtiques i Informàtica, Universitat de Barcelona, Barcelona, Spain
e-mail: vnavas@crm.cat

Á. González
Centre de Recerca Matemàtica, Bellaterra, Barcelona, Spain

GFZ German Research Centre for Geosciences, Potsdam, Germany

I. Serra
Centre de Recerca Matemàtica, Bellaterra, Barcelona, Spain

Barcelona Supercomputing Center, Barcelona, Spain

E. Vives
Departament de Matèria Condensada, Facultat de Física, Universitat de Barcelona, Barcelona, Catalonia, Spain

Universitat de Barcelona Institute of Complex Systems (UBICS), Facultat de Física, Universitat de Barcelona, Barcelona, Catalonia, Spain

Á. Corral
Barcelona Graduate School of Mathematics, Barcelona, Spain

Complexity Science Hub Vienna, Vienna, Austria

Departament de Matemàtiques, Universitat Autònoma de Barcelona, Barcelona, Spain

Centre de Recerca Matemàtica, Bellaterra, Barcelona, Spain

F. Font, T. G. Myers (eds.), *Multidisciplinary Mathematical Modelling*,
SEMA SIMAI Springer Series 11, https://doi.org/10.1007/978-3-030-64272-3_5

likelihood estimation, this methodology provides rigorous statistical information to discern whether power-law exponents characterizing different datasets can be considered equal to each other or not. This procedure is applied to the Gutenberg–Richter law for earthquakes and for synthetic earthquakes (acoustic emission events) generated in the laboratory: labquakes (Navas-Portella et al. Phys Rev E 100:062106, 2019).

1 Introduction

Generally speaking, a complex system can be understood as a large number of interacting elements whose global behavior cannot be derived from the local laws that characterize each of its components. The global response of the system can be observed at different scales, and the vast number of degrees of freedom hampers the prediction of the system dynamics. In this context, a probabilistic description of the phenomenon is needed in order to reasonably characterize it in terms of random variables. When the response of these systems exhibits lack of characteristic scales, it can be described in terms of power-law-type probability density functions (PDF), $f(x) \propto x^{-\gamma}$, where x corresponds to the values that can be taken by the random variable that characterizes the response the system can take, $\propto$ denotes the proportionality, and the power-law exponent γ acquires values larger than one. The power law is the only function that is invariant under any scale transformation of the variable x [1]. This property of scale invariance confers a description of the response of the system where there are no characteristic scales. This common framework is very usual in different disciplines [2, 3] such as condensed matter physics [4], economics [5], linguistics [6], geoscience [7], and, in particular, seismology [8, 9].

It has been broadly studied [10, 11] that different complex systems can be grouped into the same universality class when they present common values of all their power-law exponents and share the same scaling functions. Therefore, it is important to determine these exponents rigorously, not only to properly characterize phenomena but also to provide a good classification into universality classes.

In practice, exponents are difficult to measure empirically. Due to experimental limitations that distort the power-law behavior, the property of scale invariance can only be measured in a limited range. Therefore, when a power-law distribution is fitted to empirical data is more convenient to talk about local or restricted scale invariance. In this context, the wider the range the fitted PDF spans, the more reliable and strong this property will be.

A paradigmatic example of power-law behavior in complex systems is the well-known Gutenberg–Richter (GR) law for earthquakes [12]. This law states that, above a lower cut-off value, earthquake magnitudes follow an exponential distribution, in terms of the magnitude PDF,

$$f(m) = (b \log 10) 10^{-b(m-m_{min})} \propto 10^{-bm}, \tag{1}$$

defined for $m \geq m_{min}$, with m the magnitude (moment magnitude in our case), m_{min} the lower cut-off in magnitude, b is the so-called $b-$value, and log corresponds to the natural logarithm. The general relationship between seismic moment x and moment magnitude m is given by

$$x = 10^{\frac{3}{2}m+9.1}, \tag{2}$$

measured in units of Nm [13, 14]. Provided that in a change of variables such as $m \to x$ the probability is invariant, the following property must be true $f_X(x)dx = f_{\mathrm{m}}(m)dm$. Considering this probability invariance and combining Eqs. (1) and (2), the GR law is a power-law distribution when it is written as a function of the seismic moment x

$$f(x) = \frac{2}{3}\frac{b}{x_{min}}\left(\frac{x}{x_{min}}\right)^{-\left(1+\frac{2}{3}b\right)} = \frac{\gamma-1}{x_{min}}\left(\frac{x}{x_{min}}\right)^{-\gamma}, \tag{3}$$

where we conveniently define $\gamma = 1+\frac{2}{3}b$ ($b > 0$ and $\gamma > 1$), and x_{min} corresponds to the value of the seismic moment of the cut-off magnitude m_{min} [15] introduced in Eq. (2). Note that this PDF has a domain $x \in [x_{min}, +\infty)$.

An earthquake catalog is an empirical dataset that characterizes each earthquake by an array of observations: time of occurrence, spatial coordinates, magnitude, etc. The magnitude m_{min} is usually associated to the completeness threshold, such as all earthquakes with $m \geq m_{min}$ are recorded in the catalog [15]. For $m < m_{min}$, some events are missing from the catalog due to the difficulties of detecting them (e.g., [16, 17]), especially when the seismic activity rate suddenly increases, such as in aftershock sequences, in which earthquake waveforms tend to overlap each other and are difficult to detect [15, 18–20]. This incompleteness distorts the power-law behavior below m_{min}, whose value should be an upper bound to encompass these variations.

One has to keep in mind that there also exists an upper cut-off due to finite-size effects [21], implying that, at a certain value of the magnitude, there are deviations from the power-law behavior. Consequently, strictly speaking, the range of validity of the GR law cannot be extended up to infinity [7, 22]. By ignoring which is the model that conveniently fits the tail of the distribution, the power-law behavior has to be restricted to an upper cut-off x_{max}. By considering a generic power-law distribution $f_X(x) = Cx^{-\gamma}$, where C is a normalizing constant and the support $x \in [x_{min}, x_{max}]$, the PDF for the truncated GR law is written as

$$f(x) = \frac{1-\gamma}{x_{max}^{1-\gamma} - x_{min}^{1-\gamma}} x^{-\gamma}, \tag{4}$$

defined for $x \in [x_{min}, x_{max}]$.

Recent studies regarding the acoustic emission (AE) in compression experiments of porous glasses and minerals [23–28] or wood [29] have focused the attention on the energy distribution of AE events due to the similarities with the GR law for earthquakes [22]. According to the terminology that is used in some of these studies, we will name as labquakes those AE events that occur during the compression of materials.

Earthquake and labquake catalogs as well as other empirical datasets in complex systems only report a limited range of events, making it difficult to estimate parameters of the power-law PDF accurately. In this work, we try to solve this problem by combining datasets with rigorous statistical tools, with the goal of finding a broader range of validity when the different datasets correspond to the same system. If different datasets can be combined and characterized by a unique power-law exponent, it means that the particular exponents of each dataset are statistically compatible. When different phenomena share the same power-law exponents for the distributions of all their observables, they can be classified into the same universality class. Consequently, this methodology represents a statistical technique to discern whether different phenomena can be classified into the same universality class or not.

In order to conveniently classify earthquakes and labquakes into the same universality class, all the observables should be taken into account and all the corresponding exponents should be compatible. In this work, we will illustrate this methodology by focusing on the distribution of seismic moment for earthquakes and the AE energy for labquakes. The results will reveal whether they are candidates to be classified into the same universality class or not, but the final establishment of their belonging to the same universality class would require the study of all the possible observables.

The manuscript is structured as follows: Sect. 2 deals with the procedure of mergin datasets. This method is first tested with synthetic data (Sect. 3). The earthquake and charcoal labquake catalogs used are presented in Sect. 4, which also explores the effect of the data resolution in the fitting procedure. Earthquake catalogs are merged in Sect. 5 and then merged with labquake catalogs in Sect. 6. The conclusions are set out in Sect. 7.

2 Merging Datasets

By considering n_{ds} datasets of N_i ($i = 1, \ldots, n_{ds}$) observations each, one wants to fit a general power-law distribution with a unique global exponent for all of them. Let us assume that for the i-th dataset, the variable $\mathcal{X}$ (seismic moment if one works with the GR law for earthquakes or AE energy if one works with the GR law for labquakes) follows a power-law PDF $f_{\mathcal{X}}(x; \gamma_i, x_{min}^{(i)}, x_{max}^{(i)})$ given by Eq. (4) from a certain lower cut-off $x_{min}^{(i)}$ to an upper cut-off $x_{max}^{(i)}$ with exponent γ_i and number of data n_i ($n_i \leq N_i$) in the range $\left[x_{min}^{(i)}, x_{max}^{(i)}\right]$. Note that one can also consider the untruncated power-law model for the i-th dataset if $x_{max}^{(i)} \to \infty$. One can state that data from the i-th dataset does not lead to the rejection of the power-law hypothesis for a certain range (see Refs. [7, 30], or alternatively, Ref. [31]). Note that, in the i-th dataset, the variable X can acquire values in a range typically spanning several orders of magnitude.

Generally, the procedure of merging datasets is performed by selecting upper and lower cut-offs $x_{min}^{(i)}$ and $x_{max}^{(i)}$ ($x_{min}^{(i)} < x_{max}^{(i)}$) for each dataset. Note that "merging" does not imply that events as a whole are grouped together but their values of the corresponding observable under study are instead lumped together to a new dataset. Data outside these ranges are not considered. All the possible combinations of cut-offs $\{x_{min}\}$ and $\{x_{max}\}$ are checked with a fixed resolution (see below). The Residual Coefficient of Variation (CV) test can be used to fix some upper cut-offs, thus reducing the computational effort. For more details about the CV test applied in this context, see Ref. [32]. Two models can be tested:

- **Model One Exponent (OneExp)**: All datasets are merged by considering a unique global exponent Γ ($\gamma_i = \Gamma$ for all datasets).
- **Model Multi Exponent (MultiExp)**: All datasets are merged, but each one with its own exponent γ_i ($i = 1, \ldots, n_{\text{ds}}$).

Note that model OneExp is nested in model MultiExp and the difference in the number of parameters characterizing these models is $n_{\mathcal{L}}^{(MultiExp)} - n_{\mathcal{L}}^{(OneExp)} = n_{ds} - 1$. Since one is interested in merging datasets with a unique global exponent (model OneExp), enough statistical evidence that this simpler model is suitable to fit the data is needed. The fit is performed by means of the following protocol:

0. **Select a given set of cut-offs** $\left[x_{min}^{(i)}, x_{max}^{(i)}\right]$, ($x_{min}^{(i)} < x_{max}^{(i)}$) for $i = 1, \ldots, n_{ds}$.
1. **Maximum likelihood estimation (MLE) of model OneExp**: The log-likelihood function of model OneExp can be written as

$$\log \mathcal{L}_{OneExp} = \sum_{i=1}^{n_{\text{ds}}} \sum_{j=1}^{n_i} \log f_X \left(x_{ij}; \Gamma, x_{min}^{(i)}, x_{max}^{(i)} \right) , \tag{5}$$

where x_{ij} corresponds to the n_i values of the variable X that are in the range $x_{min}^{(i)} \leq x_{ij} \leq x_{max}^{(i)}$ in the i-th dataset, log is the natural logarithm, and Γ is the global exponent. The definition of likelihood is consistent with the fact that likelihoods from different datasets can be combined in this way [33, p. 27]. At this step, one has to find the value $\hat{\Gamma}$ of the global exponent Γ that maximizes the log-likelihood expression in Eq. (5). For the particular expressions corresponding to the untruncated and truncated power-law PDF, see Eqs. (3) and (4). If all the power-law distributions are untruncated, this exponent can be easily found analytically [34] as

$$\hat{\Gamma} = 1 + \frac{\sum_{i=1}^{n_{ds}} n_i}{\sum_{i=1}^{n_{ds}} \frac{n_i}{\hat{\gamma}_i - 1}} ,$$

where the hats denote the values of the exponents that maximize the log-likelihood of the particular power-law distribution (model MultiExp) and the general one in Eq. (5). See Ref. [34] for more details about the relationship between the global and particular exponents. If truncated power-law distributions are considered, one has to use a numerical method in order to determine the exponent $\hat{\Gamma}$ that maximizes this expression [30, 35].

2. **MLE of model MultiExp**: The log-likelihood function of the model MultiExp can be written as

$$\log \mathcal{L}_{MultiExp} = \sum_{i=1}^{n_{\text{ds}}} \sum_{j=1}^{n_i} \log f_X\left(x_{ij}; \gamma_i, x_{min}^{(i)}, x_{max}^{(i)}\right), \tag{6}$$

using the same notation as in Eq. (5). For the particular expressions corresponding to the truncated and untruncated power-law PDFs, see Eqs. (4) and (3) in Sect. 2. The values of the exponents that maximize Eq. (6) are denoted as $\hat{\gamma}_i$.

3. **Likelihood ratio test**: The likelihood ratio test (LRT) for the models OneExp and MultiExp is used to check whether the model OneExp is good enough to fit the data or not in comparison with the model MultiExp. For more details about the LRT applied in this context, see Ref. [32]. If the model OneExp is not rejected, the procedure goes to step (4). Otherwise, this fit is discarded and the procedure goes back to step (0). Note that the model OneExp can be a good model to fit if the particular exponents $\hat{\gamma}_i$ do not exhibit large differences among each other in relation to their uncertainty.

4. **Goodness-of-fit test**: In order to check whether it is reasonable to consider the model OneExp as a good candidate to fit the data, the next null hypothesis is formulated H_0: the variable $\mathcal{X}$ is power-law distributed with the global exponent $\hat{\Gamma}$ for all the datasets. Two different statistics used in order to carry out the goodness-of-fit tests are used: the Kolmogorov–Smirnov Distance of the Merged Datasets (KSDMD) and the Composite Kolmogorov–Smirnov Distance (CKSD). The KSDMD statistic can be used as long as datasets overlap each other, whereas the CKSD statistic does not require this condition. For more details about how these statistics are defined and how the p-value of the test is found, see Ref. [32]. If the resulting p-value is greater than a threshold value p_c (in the present work, the thresholds $p_c = 0.05$ and $p_c = 0.20$ are used), this is considered as a valid fit and it can be stated that the variable $\mathcal{X}$ is power law distributed with exponent $\hat{\Gamma}$ along all the different datasets for the different ranges $\{x_{min}\}$ and $\{x_{max}\}$. Otherwise, this fit is not be considered as valid and the procedure goes back to step (0).

When all the combinations of cut-offs have been checked, one may have a list of valid fits. In order to determine which of them is considered the definitive, the following procedure is carried out:

1. The fit that covers the largest sum of orders of magnitude $\max\left[\sum_{i=1}^{n_{ds}} \log_{10}\left(\frac{x_{max}^{(i)}}{x_{min}^{(i)}}\right)\right]$ is chosen. If the power-law fit is untruncated, $x_{max}^{(i)}$ can be substituted by the maximum observed value $x_{top}^{(i)}$. If there is a unique candidate with a maximum number of orders of magnitude, then this is considered as the definitive global fit. Otherwise, the procedure goes to the next step.
2. The fit with the broadest global range $\max\left[\frac{\max\left(x_{max}^{(i)}\right)}{\min\left(x_{min}^{(i)}\right)}\right]$ for $i = 1, \ldots, n_{ds}$ is chosen. If there is a unique candidate, this is considered as the definitive global fit. Otherwise, the procedure goes to the next step.
3. The fit with the maximum number of data $\mathcal{N} = \sum_{i=1}^{n_{ds}} n_i$ is considered as the definitive global fit.

By means of these three steps, a unique fit has been found for all the datasets analyzed in this work. Nevertheless, one could deal with datasets in which more conditions are needed in order to choose a definitive fit unambiguously. In this case, the protocol formally concludes either by considering a subset of valid fits or choosing one of them according to the researcher's criteria. At the end of this procedure, if a unique solution is found, one is able to state that the datasets that conform the global fit correspond to phenomena that are candidates to be classified into the same universality class, at least regarding the observable $\mathcal{X}$. If no combination of cut-offs is found to give a good fit, then it can be said that there exists at least one catalog that corresponds to a phenomenon that must be classified in a different universality class.

3 Performance Over Synthetic Datasets

Once the methodology for merging datasets has been presented together with the different goodness-of-fit tests, it is important to check the performance of the method over synthetic data. In order to carry out this analysis, two untruncated power-law-distributed datasets with exponents γ_1 and γ_2, lower cut-offs $x_{min}^{(1)}$ and $x_{min}^{(2)}$, and sizes n_1 and n_2 are generated. In order to simplify the analysis, the sizes of both datasets are considered to be equal, $n_1 = n_2$. The global exponent $\hat{\Gamma}$ of the merged catalogs is estimated according to the methodology explained in the main text, and the LRT statistic $2\mathcal{R}_e$ is computed according to the methods presented in Sect. 2. Given that the difference on parameters between model 1 and 2 in this case is 1, the critical value of the test with a level of risk equal to 0.05 is $2\mathcal{R}_c = 3.84$. If the empirical LRT statistic is found to be larger than the critical one, one then rejects the null hypothesis that the simpler model 1 is good enough to describe data and, consequently, more parameters are needed. Once the global exponent and the likelihood ratio statistic are found, the two different goodness-of-fit tests explained in Ref. [32] are performed.

The analysis is performed for different dataset sizes as well as different power-law exponents γ_1 and γ_2. The results of the method for synthetic datasets are shown in Table 1. Five groups are presented depending on the relative difference δ_γ between exponents. The p-values of the CKSD and KSDMD goodness-of-fit tests have been computed with 10^4 Monte Carlo simulations. In order to compare our results with a test that checks whether two power-law exponents are significantly different or not, the p_{norm}-value of the z-test detailed in Ref. [9] is also shown. These p-values can be computed by assuming that, for a sufficiently large sample size, the z-statistic follows a normal distribution with zero mean and standard deviation equal to one. As one would expect, if both datasets have exactly the same power-law exponent, the null hypothesis that variable $\mathcal{X}$ is power-law-distributed for all the datasets with the global exponent $\hat{\Gamma}$ is not rejected. When the CKSD goodness-of-fit test rejects the null hypothesis of a power-law distribution with a global exponent, the z-test also rejects the null hypothesis of considering $\gamma_1 = \gamma_2$. The same does not apply for the KSDMD goodness-of-fit test, where some fits yield to non-rejectable p-values, whereas the z-test clearly rejects the null hypothesis. In this sense, one can consider that the KSDMD statistic is less strict than the CKSD statistic. However, for the sample sizes that are involved in this work, both goodness-of-fit tests reject the null hypothesis for sufficiently large difference in the exponents. It can also be seen that the null hypothesis is rejected independently on the goodness-of-fit test for those fits in which the Likelihood Ratio statistic exceeds the critical value $2\mathcal{R}_c = 3.84$. This fact justifies the decision of performing the LRT before the goodness-of-fit test.

4 Earthquake and Charcoal Labquake Catalogs

We have selected catalogs that have different completeness magnitudes in order to cover different magnitude ranges. We hope that a convenient combination of these datasets will give us a larger range of validity of the GR law with a unique exponent. Let us briefly describe the catalogs that have been used in this work (see also Fig. 1):

- Global Centroid Moment Tensor (CMT) Catalog: It comprises earthquakes worldwide since 1977 [36, 37], from which we analyze the dataset until the end of 2017. This catalog reports the values of the moment magnitude as well as the seismic moment. Given that the seismic moment is provided with three significant digits, the resolution of the magnitude in the catalog is approximately $\Delta m \simeq 10^{-3}$.
- Yang–Hauksson–Shearer (YHS) A: It records earthquakes in Southern California with $m \geq 0$ in the period 1981–2010 [38]. This catalog does not report the seismic moment but a preferred magnitude that is approximately converted into seismic moment according to Eq. (2). The resolution of the catalog is $\Delta m = 0.01$.

Table 1 Performance of the methodology for merging datasets explained in the main text for two untruncated power laws with exponents γ_1 and γ_2, lower cut-offs $x_{min}^{(1)} = 10^{12}$ and $x_{min}^{(2)} = 10^{14}$ (arbitrary units), and sizes n_1 and n_2. σ_1 and σ_2 correspond to the standard deviations of the MLEs. Five different groups are presented depending on the relative difference of the power-law exponents δ_γ. For these datasets, the fitted global exponent $\hat{\Gamma}$ is estimated and the LRT statistic $2\mathcal{R}_e$ is computed. The statistics $D_e^{(CKSD)}$ and $D_e^{(KSDMD)}$ together with their corresponding p-values p_{value}^{CKSD} and p_{value}^{KSDMD} are presented for each combination of datasets. As a complement, the z-statistic and the p-value p_{norm}-z from to the z-test are also shown

n_1	γ_1	σ_1	n_2	γ_2	σ_2	δ_γ (%)	$\hat{\Gamma}$	$2\mathcal{R}_e$	$D_e^{(CKSD)}$	p_{value}^{CKSD}	$D_e^{(KSDMD)}$	p_{value}^{KSDMD}	z	p_{norm}-z
10^2	1.50	0.05	10^2	1.50	0.05	0	1.471	0.624	1.635	0.425	0.075	0.453	0.788	0.431
10^3	1.50	0.02	10^3	1.50	0.02	0	1.503	1.284	1.316	0.801	0.023	0.595	1.132	0.257
10^4	1.50	0.005	10^4	1.50	0.005	0	1.501	0.139	1.289	0.837	0.009	0.986	0.373	0.707
10^2	1.53	0.05	10^2	1.50	0.05	2	1.484	1.443	1.660	0.395	0.070	0.578	1.195	0.232
10^3	1.53	0.02	10^3	1.50	0.02	2	1.518	0.029	1.239	0.876	0.017	0.893	0.170	0.865
10^4	1.53	0.005	10^4	1.50	0.005	2	1.516	14.037	3.187	5×10^{-4}	0.013	0.024	3.745	1.81×10^{-4}
10^2	1.575	0.058	10^2	1.50	0.05	5	1.514	3.154	1.815	0.249	0.058	0.755	1.755	0.079
10^3	1.575	0.018	10^3	1.50	0.02	5	1.541	2.458	2.032	0.111	0.031	0.180	1.567	0.117
10^4	1.575	0.006	10^4	1.50	0.005	5	1.536	64.957	4.740	0	0.029	0	9.474	0
10^2	1.65	0.065	10^2	1.50	0.05	10	1.528	6.956	2.262	0.046	0.054	0.832	2.571	0.010
10^3	1.65	0.02	10^3	1.50	0.02	10	1.57	22.365	3.369	10^{-4}	0.044	0.015	4.690	2.73×10^{-6}
10^4	1.650	0.007	10^4	1.50	0.005	10	1.566	329.552	10.601	0	0.053	0	17.933	0
10^2	1.725	0.073	10^2	1.50	0.05	15	1.551	11.566	2.661	0.008	0.090	0.260	3.262	0.001
10^3	1.725	0.023	10^3	1.50	0.02	15	1.598	51.266	4.608	0	0.066	10^{-4}	7.026	2.12×10^{-12}
10^4	1.725	0.007	10^4	1.50	0.005	15	1.593	667.096	14.507	0	0.075	0	25.204	0

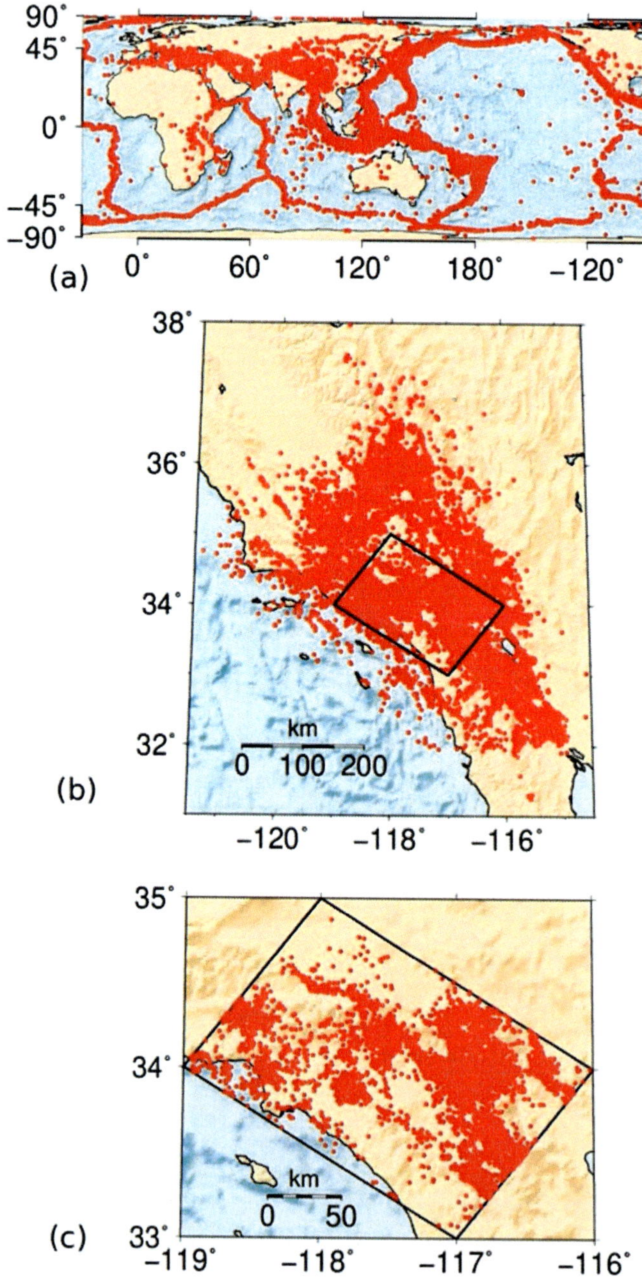

Fig. 1 Earthquake epicenters of the different catalogs. Horizontal and vertical axes correspond to longitude and latitude in degrees, respectively. (**a**) Full CMT catalog for the period 1977–2017 [36, 37]. (**b**) Entire YHS A Catalog [38] of Southern California for the period 1981–2010. (**c**) Subcatalog corresponding to a region (YHS B) of the YHS catalog for Los Angeles area for the period 2000–2010. Maps in cylindrical equal-area projection (top) and sinusoidal projection (middle and bottom), produced with Generic Mapping Tools [39]

- Yang–Hauksson–Shearer (YHS) B: It is a subset of YHS A that contains the earthquakes in the region of Los Angeles in the period 2000–2010 [38]. This LA region is defined by the following four vertices in longitude and latitude: $(119°W, 34°N)$,$(118°W, 35°N)$,$(116°W, 34°N)$, and $(117°W, 33°N)$ (see Fig. 1). This region has been selected because it is among the best monitorized ones [17, 40]. Furthermore, we selected this time period due to the better detection of small earthquakes than in previous years [40], which should reduce the completeness magnitude of the catalog [17]. The magnitude resolution of this catalog is the same as YHS A.

Figure 1 shows the epicentral locations of the earthquakes contained in each catalog. In the statistical analysis, in order to not to count the same earthquake more than once, the spatio-temporal window corresponding to the YHS B catalog has been excluded from the YHS A and the spatio-temporal window corresponding to the YSH A catalog has been excluded from the CMT catalog.

We performed one uniaxial compression experiment of charcoal in a conventional test machine Z005 (Zwick/Roell). Samples with no lateral confinement were placed between two plates that approached each other at a certain constant displacement rate $\dot{z}$. Simultaneous to the compression, recording of an AE signal was performed by using a piezoelectric transducer embedded in one of the compression plates. The electric signal $U(t)$ was pre-amplified, band filtered (between 20 kHz and 2 MHz), and analyzed by means of a PCI-2 acquisition system from Euro Physical Acoustics (Mistras Group) with an AD card working at 40 megasamples per second with 18 bits precision [41]. Signal pre-amplification was necessary to record small AE events. Some values of the pre-amplified signal were so large that could not be detected correctly by the acquisition system. This fact led to signal saturation and, consequently, an underestimated energy of the AE event [34]. Recording of data stopped when a big failure event occurred, and the sample got destroyed.

An AE event (often called AE hit in specialized AE literature) starts at the time t_j when the signal $U(t)$ exceeds a fixed detection threshold and finishes at time $t_j + \tau_j$ when the signal remains below threshold from $t_j + \tau_j$ to at least $t_j + \tau_j +$ 200 µs. The energy E_j of each event is determined as $E_j = \frac{1}{R}\int_{t_j}^{t_j+\tau_j} U^2(t)dt$, where R is a reference resistance of 10 kΩ. This AE energy corresponds to the radiated energy received by the transducer. At the end of an experiment, a catalog of events is collected, each of them characterized by a time of occurrence t_j, energy E_j, and duration τ_j.

4.1 *Effect of the Magnitude Resolution in Earthquake Catalogs*

A different issue that can be addressed is whether the magnitude resolution in earthquake catalogs affects the fitting procedures when PDFs are considered as continuous. As explained in Sect. 4, some catalogs provide the magnitudes with a given resolution Δm, leading to discretized values when they are transformed into seismic moments in order to fit a power-law distribution. In order to check the effect of the resolution Δm when fitting a power law, n exponentially distributed magnitudes were generated according to the Gutenberg–Richter law (assuming $b = 1$ and $m_{min} = 3$) from uniform random numbers u according to

$$m = m_{min} - \frac{\log_{10}(1-u)}{b} .$$

Once generated, these values were binned according to the bin width or resolution Δm and converted into seismic moment according to Eq. (2). The fitting procedure explained in Sect. 2 was then applied, and the results for different sample sizes ($n = 10^2$, $n = 10^3$ and $n = 10^4$) of magnitudes sampled from a Gutenberg–Richter law (b-value= 1 and $m_{min} = 3$) and discretized according to different values of the resolution Δm are shown in Table 2.

In Fig. 2, the fitted $\hat{b}$-value as a function of the resolution Δm is shown for the three different sample sizes. As it can be observed, the value of the maximum

Table 2 Results of simulating magnitudes sampled from a Gutenberg–Richter law with $m_{min} = 3$ and b-value= 1 for different sample sizes n and different resolutions Δm. The fitted $\hat{b}$-values and their standard deviations σ were computed by means of MLE, and p-values were extracted from the KS goodness-of-fit test explained in Ref. [30]. Data are plotted in Fig. 2

Δm	m_{min}	b-value	n	$\hat{b}$	σ	p_{value}
1	3	1	10^2	3.948	0.394	0
			10^3	4.4326	0.140	0
			10^4	4.024	0.040	0
0.5	3	1	10^2	1.448	0.145	0
			10^3	1.876	0.059	0
			10^4	1.840	0.018	0
0.1	3	1	10^2	1.070	0.106	0
			10^3	1.113	0.035	0
			10^3	1.122	0.011	0
0.05	3	1	10^2	0.994	0.099	0.022 ± 0.001
			10^3	1.060	0.033	0
			10^4	1.086	0.011	0
0.01	3	1	10^2	1.050	0.105	0.578 ± 0.004
			10^3	1.013	0.032	0.267 ± 0.004
			10^4	0.997	0.010	0
0.005	3	1	10^2	1.076	0.108	0.856 ± 0.004
			10^3	1.039	0.033	0.396 ± 0.005
			10^4	0.995	0.010	0.033 ± 0.002
0.001	3	1	10^2	1.067	0.102	0.588 ± 0.005
			10^3	1.018	0.0322	0.828 ± 0.004
			10^4	1.009	0.010	0.386 ± 0.005

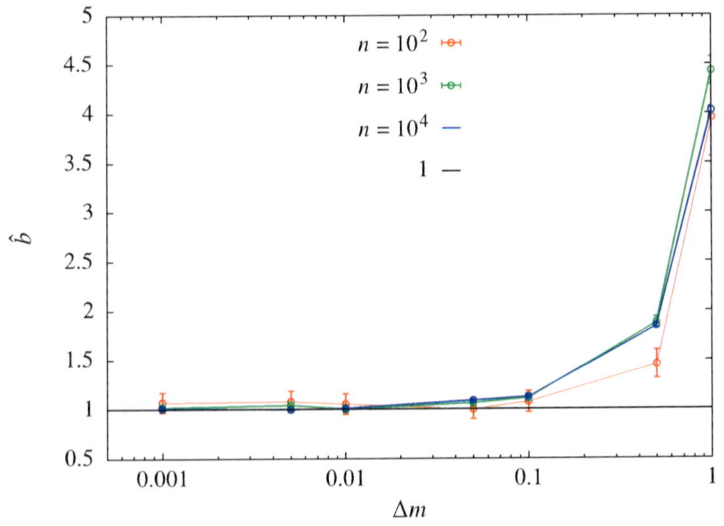

Fig. 2 Maximum likelihood estimator $\hat{b}$-value as a function of the resolution in magnitudes Δm for different sample sizes. Data are sampled from a Gutenberg–Richter law with $m_{min} = 3$ and b-value$= 1$. The horizontal black line corresponds to the real b-value. Data are shown in Table 2

likelihood (ML) estimator converges to the expected b-value$= 1$ as the magnitude bin width is decreased. The ML estimator is biased for resolutions larger than 0.1 and is acceptable, within the error bars, for smaller values of the resolution [42]. However, the results of the goodness-of-fit test lead to rejectable p-values for large samples even when the ML estimator is unbiased (see Table 2). A resolution of $\Delta m = 10^{-3}$ ensures that the largest sample of $n = 10^4$ yields non-rejectable p-values. Some results presented in this chapter deal with catalogs whose resolution and number of events could result in rejectable fits. However, non-rejectable p-values were found instead, leading to the conclusion that data in those real catalogs follow a power-law distribution even more reliably.

5 Merging Earthquake Catalogs

The GR law states that the seismic moment can be considered as a random variable $\mathcal{M}$ that is power-law distributed. In its usual form, the GR law fits a power-law model that contemplates a unique power-law exponent. Nevertheless, several studies have elucidated the existence of a double-power-law behavior in the GR law for global seismicity [7, 43, 44]. Corral and Gonzalez [7] pointed out that a truncated power law with exponent $\gamma \simeq 1.66$ cannot be rejected up to $m_{max} \simeq 7.4$ and a second power-law tail emerges from $m'_{min} = 7.67$ with an exponent $\gamma'_M = 2.1 \pm 0.1$. Furthermore, by fixing the upper truncation at $m_{max} = m'_{min} = 7.67$, the truncated power-law hypothesis cannot be rejected (see Table 3). Consequently, if one wishes to fit a power-law PDF with a unique exponent, all those earthquakes with $m \geq m_{max} = 7.67$ should be excluded. For the CMT catalog, the upper cut-off is fixed at $M_{max} = 10^{\frac{3}{2}m_{max}+9.1}$, whereas the other catalogs can be safely fitted by untruncated power-law PDFs because the CV test does not reject the hypothesis of a unique power-law tail and the magnitudes that are studied are considerably smaller than those in the CMT catalog (see Table 3).

Therefore, two untruncated power-law distributions and a third one that is truncated for the CMT catalog are considered in order to merge catalogs. For each decade, 5 values of $M^{(i)}_{min}$ equally spaced in logarithmic scale are sampled, and all the possible combinations of cut-offs $M^{(1)}_{min}$, $M^{(2)}_{min}$, and $M^{(3)}_{min}$ are checked for a fixed upper truncation $M^{(3)}_{max}$. The labels (1), (2), and (3) correspond to the catalogs YSH B, YSH A, and CMT, respectively.

In Table 4, the results of the global fit for models OneExp and MultiExp are shown. The same global fit is found independently from the choice of the test statistic used in the goodness-of-fit test. A b-value very close to one holds for more than 8 orders of magnitude in seismic moment from $m_{min} = 1.93$ to $m_{max} = 7.67$ (see Fig.3). The value of the global exponent is approximately in agreement with the harmonic mean of the particular exponents of the GR law for each catalog [34, 45]. Due to the upper truncation, the value of the global exponent is not exactly the same as the value of the harmonic mean of the particular exponents. The results for the CKSD statistic do not show remarkable differences if the critical p-value is set to $p_c = 0.20$ (see Table 4). The definitive fit is the one whose goodness-of-fit test has been performed with a threshold value of $p_c = 0.20$.

This is the broadest fitting range that has been found for the Gutenberg–Richter law for earthquakes with a unique value of the exponent [9]. There are catalogs of tiny mining-induced earthquakes that exhibit a much smaller completeness magnitude [46] than the ones of natural seismicity used in this work. They were not considered here because they are not currently public and show b-values significantly different [47] from the one found here, which would result in non-acceptable fits when merging them with the other catalogs, possibly pointing to a different universality class.

6 Merging Earthquake and Labquake Catalogs: Universality

Motivated by the fact that the power-law exponents of earthquake catalogs and charcoal labquakes are very similar, the methodology for merging datasets explained in Sect. 2 is here applied by adding also the catalog of charcoal labquakes.

Table 3 Results of fitting the GR law for each individual catalog. The total number of earthquakes in each catalog is given by N, whereas the number of data entering into the fit is n. The value m_{top} corresponds to the maximum observed value for each catalog. The GR law is valid for each catalog from $[m_{min}, m_{max}]$ with a particular b-value. m_{max} has no upper limit for any of the fits except for the CMT catalog, in which $m_{max}^{(3)} = 7.67$ (so $M_{max}^{(3)} = 4.03 \times 10^{20}$ Nm). Numbers in parentheses correspond to the error bars estimated with one σ in the scale given by the last digit. The p-value of the fits has been computed with 10^3 simulations and $p_c = 0.05$

	N	n	m_{min}	M_{min} (Nm)	m_{top}	M_{top}(Nm)	$\hat{b}$-value	$\hat{\gamma}_M$	p_{fit}
(1) YHS B	26330	3412	1.93	10^{12}	5.39	1.53×10^{17}	0.99(3)	1.66(2)	0.072(8)
(2) YHS A	152924	4353	3.17	7.08×10^{13}	7.20	7.94×10^{19}	0.98(1)	1.65(1)	0.080(9)
(3) CMT	48637	22336	5.33	1.24×10^{17}	9.08	5.25×10^{22}	0.982(7)	1.655(5)	0.36(2)

Table 4 Results of fitting the models OneExp and MultiExp to the earthquake catalogs when performing the goodness-of-fit test with the CKSD statistics and different values of p_c. If the goodness-of-fit test is performed with the KSDMD statistic, the same set of cut-offs as the fit done by using CKSD$_{0.20}$ is found. Same symbols as in Table 3. OM corresponds to the orders of magnitude, and $\hat{b}_g$-value is the global b-value$= \frac{3}{2}(\Gamma_M - 1)$. The value $M_{max}^{(i)}$ is replaced for $M_{top}^{(i)}$ for untruncated fits. Note that $7.67 - 1.93 = 5.74$ units in magnitude correspond to 8.6 orders of magnitude in seismic moment

Model MultiExp	n	M_{min}	M_{min} (Nm)	OM	$\hat{b}$-value	$\hat{\gamma}_M$	p_{fit}
(1) YHS B	3412	1.93	10^{12}	5.18	0.99(1)	1.633(7)	0.072(8)
(2) YHS A	3500	3.27	10^{14}	5.90	0.99(2)	1.66(1)	0.089(9)
(3) CMT	19,003	5.40	1.58×10^{17}	3.40	0.98(8)	1.655(5)	0.26(1)
Model OneExp	N			$\sum$OM	$\hat{b}_g$-value	$\hat{\Gamma}_M$	p_{fit}
$p_c = 0.05$	25,915	1.93	10^{12}	14.48	0.991(6)	1.661(4)	0.079(9)
Model MultiExp	n	M_{min}	M_{min} (Nm)	OM	$\hat{b}$-value	$\hat{\gamma}_M$	p_{fit}
(1) YHS B	3412	1.93	10^{12}	5.18	0.99(1)	1.633(7)	0.072(8)
(2) YHS A	3500	3.27	10^{14}	5.90	0.99(2)	1.66(1)	0.089(9)
(3) CMT	10,422	5.67	3.98×10^{17}	3	1.00(1)	1.663(7)	0.62(2)
Model OneExp	N			$\sum$OM	$\hat{b}_g$-value	$\hat{\Gamma}_M$	p_{fit}
$p_c = 0.20$	17,334	1.93	10^{12}	14.08	1.000(8)	1.667(5)	0.326(5)

It is important to stress the fact that the seismic moment does not correspond with the radiated energy E_r by the earthquake, which would be the reasonable energy to compare with the AE energy. For this study, the ratio of seismically radiated energy over the seismic moment is considered as constant so that the values of the seismic moment should just be multiplied by a unique factor. The value of this unique factor is $\frac{E_r}{M} = 10^{-4.6}$ [14], where M is the seismic moment. In this case, E_r corresponds to the energy radiated in seismic waves by earthquakes and the AE energy.

It can be shown that, for both models OneExp and MultiExp, multiplying the variable by a constant factor only introduces a constant term in the log-likelihood that does not change neither its maximum nor the difference of the log-likelihoods. Let us suppose n_{ds} datasets and that data in the i-th dataset is described, for simplicity and without losing generality, in terms of untruncated power-law PDFs. The log-likelihood function for both OneExp and MultiExp models are, respectively,

$$\begin{aligned} \log \mathcal{L}_{OneExp} &= \sum_{i=1}^{n_{ds}} \sum_{j=1}^{n_i} \log f_X\left(x_{ij}\right) \\ &= \sum_{i=1}^{n_{ds}} \sum_{j=1}^{n_i} \log \left(\frac{\Gamma - 1}{x_{min}^{(i)1-\Gamma}} x_{ij}^{-\Gamma} \right). \end{aligned} \tag{7}$$

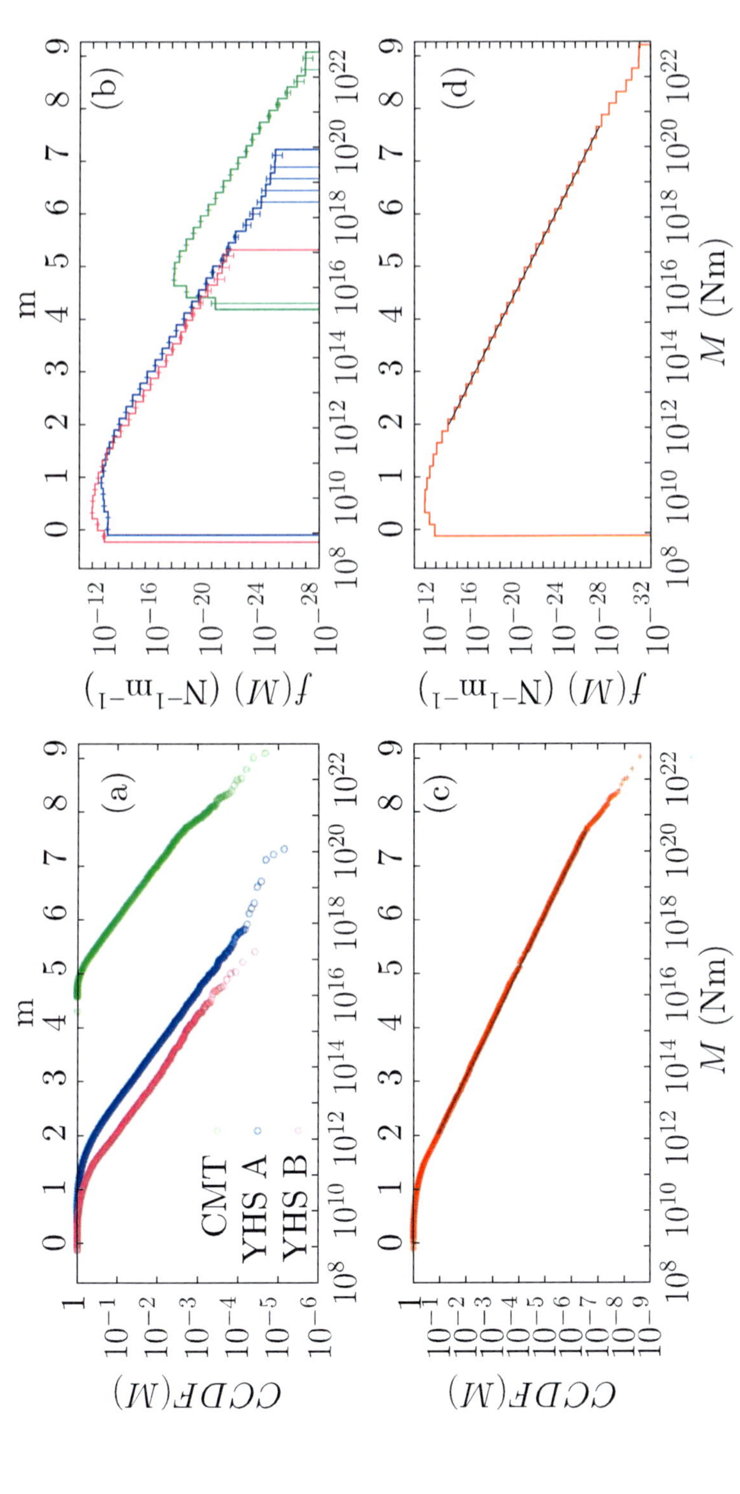

Fig. 3 Estimated Complementary Cumulative Distribution Functions (CCDF) of the Gutenberg–Richter law for each earthquake catalog (**a**) and for the merged catalogs (**c**). Estimated PDFs $f(M)$ of the Gutenberg–Richter law for each catalog (**b**) and for the merged catalogs (**d**). The merged histogram is plotted by following the procedure explained in Ref. [34]. Fits are represented by solid black lines. Top axis represents the same scale in moment magnitude

$$\log \mathcal{L}_{MultiExp} = \sum_{i=1}^{n_{ds}} \sum_{j=1}^{n_i} \log f_X\left(x_{ij}\right) = \sum_{i=1}^{n_{ds}} \sum_{j=1}^{n_i} \log\left(\frac{\gamma_i - 1}{x_{min}^{(i)1-\gamma_i}} x_{ij}^{-\gamma_i}\right) . \tag{8}$$

Let us now suppose that the values taken by the variable X in the i-th dataset are multiplied by a certain factor λ_i as well as the lower cut-off $x_{min}^{(i)}$

$$x_{ij} \rightarrow \lambda_i x_{ij} ,$$

$$x_{min}^{(i)} \rightarrow \lambda_i x_{min}^{(i)} .$$

Note that this is not strictly speaking a scale transformation since not only the variable but some parameters characterizing the function are changed. The likelihoods of the transformed data $\mathcal{L}^T$ for both models are

$$\log \mathcal{L}_{OneExp}^{T} = \sum_{i=1}^{n_{ds}} \sum_{j=1}^{n_i} \log\left(\frac{\Gamma - 1}{\left(\lambda_i x_{min}^{(i)}\right)^{1-\Gamma}} \left(\lambda_i x_{ij}\right)^{-\Gamma}\right) = \log \mathcal{L}_{OneExp} - \sum_{i=1}^{n_{ds}} n_i \log \lambda_i , \tag{9}$$

$$\log \mathcal{L}_{MultiExp}^{T} = \sum_{i=1}^{n_{ds}} \sum_{j=1}^{n_i} \log\left(\frac{\gamma_i - 1}{\left(\lambda_i x_{min}^{(i)}\right)^{1-\gamma_i}} \left(\lambda_i x_{ij}\right)^{-\gamma_i}\right) = \log \mathcal{L}_{MultiExp} - \sum_{i=1}^{n_{ds}} n_i \log \lambda_i . \tag{10}$$

The transformed likelihoods can be expressed in terms of the likelihood functions of the original data together with an additive constant term that does not depend neither on the exponent nor the cut-off. This means that the maximum of the transformed log-likelihood is attained at exactly the same value as the maximum of the original one. Given that the likelihood ratio statistic is a subtraction of the likelihood of both models, the relation between the original statistic $2\mathcal{R}$ and the one corresponding to transformed likelihood functions is

$$\begin{aligned} 2\mathcal{R}^T &= 2\left(\log \mathcal{L}^T_{MultiExp} - \log \mathcal{L}^T_{OneExp}\right) \\ &= 2\left(\log \mathcal{L}_{MultiExp} - \log \mathcal{L}_{OneExp}\right) = 2\mathcal{R}\,. \end{aligned} \tag{11}$$

Since the two additional terms in the transformed likelihoods are the same for both models, they cancel and the statistic for the likelihood ratio test is invariant under this transformation. As the CKSD statistic is a weighted average of the particular KS distances of each dataset, it does not change either. Therefore, the results shown in Table 3 would not change except for the values of the cut-offs.

The CV test does not reject the hypothesis of a unique power-law tail for the charcoal labquake catalog, and an untruncated power-law model is considered for this catalog [32]. Therefore, three untruncated power laws are fitted for the charcoal and YSH B and YSH A catalogs, and a truncated power law is fitted to the CMT catalog. For each decade, 5 different values of $x^{(i)}_{min}$ equally spaced in logarithmic scale, for a fixed upper truncation $x^{(3)}_{max}$, were checked, and all the possible combinations of cut-offs $x^{(0)}_{min}$, $x^{(1)}_{min}$, $x^{(2)}_{min}$, and $x^{(3)}_{min}$ are checked for a fixed upper truncation $x^{(3)}_{max}$. The labels (0), (1), (2), and (3) correspond to the catalogs of the charcoal experiment, YSH B, YSH A, and CMT, respectively.

The results of the global fit for the CKSD statistic are presented in Table 5. In this case, not all the catalogs overlap each other and the CKSD statistic is the only one that can be used for the goodness-of-fit test. The value of the global exponent is approximately in agreement with the harmonic mean of the particular exponents of

Table 5 Results of fitting models OneExp and MultiExp to the charcoal labquake and earthquake datasets for two different values of p_c. Same notation as in previous tables. Note that E_r represents the radiated energy in seismic waves by earthquakes and the radiated AE energy (Nm = Joule)

Model MultiExp	n	m_{min}	$E_{r,min}$ (Nm)	OM	$\hat{b}$-value	$\hat{\gamma}_{E_r}$	p_{fit}
(0) Charcoal	15,906	–	6.31×10^{-18}	6.47	0.988(8)	1.658(5)	0.15(1)
(1) YHS B	1353	2.33	9.99×10^{7}	4.58	0.98(3)	1.66(2)	0.10(1)
(2) YHS A	234	4.47	1.59×10^{11}	4.10	0.98(6)	1.65(4)	0.62(2)
(3) CMT	7689	5.80	1.59×10^{13}	2.80	1.00(1)	1.667(9)	0.393(5)
Model OneExp	N			$\sum$OM	$\hat{b}_g$-value	$\hat{\Gamma}_{E_r}$	p_{fit}
$p_c = 0.05$	25182		6.3×10^{-18}	17.95	1.003(6)	1.669(4)	0.057(7)
Model MultiExp	n	m_{min}	$E_{r,min}$ (Nm)	OM	$\hat{b}$-value	$\hat{\gamma}_{E_r}$	p_{fit}
(0) Charcoal	3555	–	6.31×10^{-17}	5.47	1.04(2)	1.69(1)	0.88(1)
(1) YHS B	1007	2.47	1.59×10^{8}	4.38	0.99(3)	1.66(2)	0.058(7)
(2) YHS A	234	4.47	1.59×10^{11}	4.10	0.98(6)	1.65(4)	0.62(2)
(3) CMT	3014	6.20	6.33×10^{13}	2.20	1.00(2)	1.67(2)	0.59(2)
Model OneExp	N			$\sum$OM	$\hat{b}_g$-value	$\hat{\Gamma}_{E_r}$	p_{fit}
$p_c = 0.20$	7810		6.3×10^{-17}	16.15	1.03(1)	1.688(8)	0.21(1)

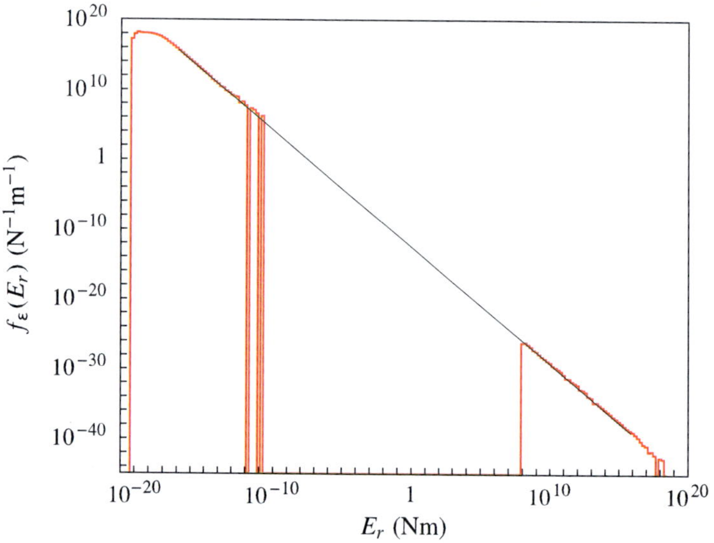

Fig. 4 Estimated PDF of the radiated energy E_r for the merged earthquake and charcoal labquake catalogs. The fit is represented by a solid black line. The methodology to construct the histogram with the three earthquake catalogs is the same as the one explained in Ref. [34], whereas the addition of the labquake histogram to this fit has been done ad hoc by conveniently rescaling both parts (those corresponding to labquakes and earthquakes, respectively). Each part has been divided by an effective number of events by assuming that the probability in each part corresponds to that obtained from a global power-law exponent with exponent $\hat{\Gamma}$ from $x_{min}^{(0)}$ to $x_{max}^{(3)}$. Note that events from the CMT, YSH A, and YSH B catalogs below their respective lower cut-offs x_{min} have been excluded in the plot

the GR law for each catalog [34]. The results do not show remarkable differences if the critical p-value is set to $p_c = 0.20$ (see Table 5). The definitive fit is the one whose goodness-of-fit test has been performed with a threshold value of $p_c = 0.20$. These results are shown in Fig. 4 and are compatible with the ones obtained from synthetic catalogs shown in Sect. 3.

Given that there is no overlapping between the charcoal labquake catalog and the earthquake ones, an alternative procedure must be applied in order to construct the histogram of the estimated global PDF. First, the three earthquake catalogs are merged according to the procedure detailed in Sect. 2. The earthquakes from the CMT catalog that are above the upper cut-off $x_{max}^{(4)}$ are also plotted, but they are not part of the global PDF. Earthquakes from the YSH B, YSH A, and CMT catalogs that are below $x_{min}^{(2)}$, $x_{min}^{(3)}$ and $x_{min}^{(4)}$, respectively, are also excluded. This piece of the histogram, which is not normalized yet, is denoted as Ea, from earthquakes. The charcoal labquake catalog histogram is plotted without normalizing by any number of events yet. The piece of histogram corresponding to charcoal labquakes is denoted as La, from labquakes. Given that no normalization has been performed yet, at this point of the procedure, the pieces (Ea) and (La) are not aligned in logarithmic scale. Consequently, the fit of a global power law with exponent $\hat{\Gamma}$ from $x_{min}^{(1)}$ to $x_{max}^{(4)}$ will not overlap with both pieces.

In order to align both pieces, one has to assume that data from $x_{min}^{(1)}$ to $x_{max}^{(4)}$ follows a power-law distribution with exponent $\hat{\Gamma}$. Given that this assumption is supported by statistical results that have been already found, it can be imposed that the probability represented by each piece corresponds to this theoretical probability. In order to achieve this imposition, the total number of counts in each piece is divided by the effective number of events in that piece

$$F\left(x = x_{top}^{(1)}; x_{min}^{(1)}, x_{max}^{(4)}, \hat{\Gamma}\right) - F\left(x = x_{min}^{(1)}; x_{min}^{(1)}, x_{max}^{(4)}, \hat{\Gamma}\right) = \frac{C\left(x_{min}^{(1)} \le x \le x_{top}^{(1)}\right)}{n_{eff}^{La}},$$

$$F\left(x = x_{max}^{(4)}; x_{min}^{(1)}, x_{max}^{(4)}, \hat{\Gamma}\right) - F\left(x = x_{min}^{(2)}; x_{min}^{(1)}, x_{max}^{(4)}, \hat{\Gamma}\right) = \frac{C\left(x_{min}^{(2)} \le x \le x_{max}^{(4)}\right)}{n_{eff}^{Ea}},$$

where $C()$ corresponds to the number of counts in the interval considered inside the parenthesis, F corresponds to the CDF of a truncated power law

$$F\left(x = x_{min}^{(1)}; x_{min}^{(1)}, x_{max}^{(4)}, \hat{\Gamma}\right) = 0\,.$$

Note that these numbers are integers for the (La) piece, but not for the (Ea) given that it has been constructed by merging several datasets and an event does not contribute necessarily one unit (see Ref. [34]). By dividing the piece (Ea) by n_{eff}^{Ea} and the piece (La) by n_{eff}^{La}, the histograms are aligned and overlapping with the theoretical fit. However, given that data outside the fit is also considered in the construction of the histogram, the integral of both pieces (Ea) and (La) will not be one and the effective area A_{eff} at this point corresponds to

$$A_{eff} = \frac{C\left(x_{low}^{(1)} \le x \le x_{top}^{(1)}\right)}{n_{eff}^{La}} + \frac{C\left(x_{min}^{(2)} \le x \le x_{top}^{(4)}\right)}{n_{eff}^{Ea}}\,.$$

In order to normalize the PDF, both pieces together with the theoretical fit must be divided by A_{eff}. Contrarily to the procedure to construct a merged histogram explained in Ref. [34], this procedure is performed ad hoc in the sense that the theoretical PDF needs to be known a priori in order to obtain a graphical representation.

Earthquake catalogs have also been merged with a charcoal labquake catalog with a global power-law exponent $\hat{\Gamma}_{E_r} = 1.688$, suggesting that these different systems would be classified into the same universality class. Further investigations involving different observables, such as the distribution of waiting times, might be necessary in order to properly classify charcoal labquakes and real earthquakes into the same universality class.

7 Conclusions

We have presented a statistical procedure to merge different datasets in order to validate the existence of universal power-law exponents across different scales or phenomena. This methodology can be useful in the study of different complex systems in order to check whether the power-law exponents obtained via maximum likelihood estimation are statistically compatible among them. Therefore, the procedure presented in this paper provides a statistical tool that enables to establish whether different complex systems can be classified into the same universality class. This work extends the results presented in [32], exploring the effect on the fits of the magnitude resolution in earthquake catalogs, and detailing the mathematical derivations.

In this work, the methodology has been applied to the Gutenberg–Richter law for earthquakes and labquakes. By merging earthquake catalogs, a global power law with a global exponent $\Gamma = 1.667$ holds for more than 8 orders of magnitude in seismic moment (from $m_{min} = 1.93$ to $m_{max} = 7.67$ in moment magnitude). To our knowledge, this is the broadest fitting range that has been found for the Gutenberg–Richter law for earthquakes with a unique value of the exponent [48]. There are catalogs of tiny mining-induced earthquakes that exhibit a much smaller completeness magnitude [46] than the ones of natural seismicity used in this work. They were not considered here because they are not currently public and show b-values significantly different [47] from the one found here, which would result in non-acceptable fits when merging them with the rest of catalogs, possibly pointing to a different universality class. Future works involving different earthquake catalogs can be carried out in order to find a broader fitting range of the Gutenberg–Richter laws and also to check whether different regions have compatible power-law exponents or not. This kind of studies would be of interest in order to statistically strengthen the geological arguments that justify the difference in the b-values observed in some regions [45].

Earthquake catalogs have also been merged with a charcoal labquake catalog with a global power-law exponent $\Gamma = 1.688$ suggesting that these different systems might be classified into the same universality class. Further investigations involving different observables, such as the distribution of waiting times, would be necessary in order to properly classify charcoal labquakes and real earthquakes into the same universality class.

Acknowledgments The research leading to these results has received funding from "La Caixa" Foundation. V. N. acknowledges financial support from the Spanish Ministry of Economy and Competitiveness (MINECO, Spain), through the "María de Maeztu" Programme for Units of Excellence in R & D (grant no. MDM-2014-0445) and the Juan de la Cierva research contract

FJCI-2016-29307 hold by Á.G.. We also acknowledge financial support from the MINECO under grant nos. FIS2015-71851-P, FIS-PGC2018-099629-B-100, and MAT2016-75823-R and from the Agència de Gestió d'Ajuts Universitaris i de Recerca (AGAUR) under grant no. 2014SGR-1307.

References

1. Christensen, K., Moloney, N.R.: Complexity and Criticality. Imperial College Press London (2005)
2. Bak, P.: How Nature Works: The Science of Self-organized Criticality. Oxford University Press, Oxford (1997)
3. Sethna, J.P., Dahmen, K.A., Myers, C.R.: Crackling noise. Nature **410**(6825), 242–250 (2001)
4. Salje, E.K.H., Dahmen, K.A.: Crackling noise in disordered materials. Annu. Rev. Cond. Matter Phys. **5**(1), 233–254 (2014)
5. Gabaix, X.: Power laws in economics: An introduction. J. Econ. Pers. **30**(1), 185–206 (2016)
6. Moreno-Sánchez, I., Font-Clos, F., Corral, A.: Large-scale analysis of Zipf's law in English texts. PLOS One **11**(1), 1–19 (2016)
7. Corral, Á., González, Á.: Power-law distributions in geoscience revisited. Earth Space Sci. **6**, 673–697 (2019)
8. Kagan, Y.Y.: Earthquakes: Models, Statistics, Testable Forecasts. Wiley, London (2013)
9. Kagan, Y.Y.: Earthquake size distribution: power-law with exponent $\beta \equiv \frac{1}{2}$? Tectonophysics **490**(1), 103–114 (2010)
10. Stanley, H.E.: Scaling, universality, and renormalization: three pillars of modern critical phenomena. Rev. Mod. Phys. **71**, S358–S366 (1999)
11. Stanley, H.E., Amaral, L.A.N., Gopikrishnan, P., Ivanov, P.Ch., Keitt, T.H., Plerou, V.: Scale invariance and universality: organizing principles in complex systems. Phys. A Stat. Mech. App. **281**(1), 60–68 (2000)
12. Utsu, T.: Representation and analysis of the earthquake size distribution: a historical review and some new approaches. Pure Appl. Geophys. **155**(2–4), 509–535 (1999)
13. Hanks, T.C., Kanamori, H.: A moment magnitude scale. J. Geophys. Res. Solid Earth **84**(B5), 2348–2350 (1979)
14. Bormann, P.: Are new data suggesting a revision of the current M_w and M_e scaling formulas? J. Seismol. **19**(4), 989–1002 (2015)
15. Woessner, J., Wiemer, S.: Assessing the quality of earthquake catalogues: estimating the magnitude of completeness and its uncertainty. Bull. Seismol. Soc. Am. **95**(2), 684–694 (2005)
16. González, Á.: The Spanish national earthquake catalogue: evolution, precision and completeness. J. Seism. **21**(3), 435–471 (2017)
17. Schorlemmer, D., Woessner, J.: Probability of detecting an earthquake. Bull. Seismol. Soc. Am. **98**(2), 2103–2117 (2008)
18. Davidsen, J., Green, A.: Are earthquake magnitudes clustered? Phys. Rev. Lett. **106**, 108502 (2011)
19. Hainzl, S.: Rate-dependent incompleteness of earthquake catalogs. Seism. Res. Lett. **87**(2A), 337–344 (2016)
20. Helmstetter, A., Kagan, Y.Y., Jackson, D.D.: Comparison of short-term and time-independent earthquake forecast models for Southern California. Bull. Seismol. Soc. Am. **96**(1), 90–106, 02 (2006)
21. Corral, Á., García-Millán, R., Moloney, N.R., Font-Clos, F.: Phase transition, scaling of moments, and order-parameter distributions in Brownian particles and branching processes with finite-size effects. Phys. Rev. E **97**, 062156 (2018)
22. Serra, I., Corral, Á.: Deviation from power law of the global seismic moment distribution. Sci. Rep. **7**(1), 40045 (2017)

23. Baró, J., Corral, Á., Illa, X., Planes, A., Salje, E.K.H. Schranz, W., Soto-Parra, D.E., Vives, E.: Statistical similarity between the compression of a porous material and earthquakes. Phys. Rev. Lett. **110**(8), 088702 (2013)
24. Nataf, G.F., Castillo-Villa, P.O., Baró, J., Illa, X., Vives, E., Planes, A., Salje, E.K.H.: Avalanches in compressed porous SiO_2-based materials. Phys. Rev. E **90**(2), 022405 (2014)
25. Navas-Portella, V., Corral, Á., Vives , E.: Avalanches and force drops in displacement-driven compression of porous glasses. Phys. Rev. E **94**(3), 033005 (2016)
26. Yoshimitsu, N., Kawakata, H., Takahashi, N.: Magnitude −7 level earthquakes: a new lower limit of self-similarity in seismic scaling relationships. Geophys. Res. Lett. **41**(13), 4495–4502 (2014)
27. Main, I.G.: Little earthquakes in the lab. Physics **6**:20 (2013)
28. Uhl, J.T., Pathak, S., Schorlemmer, D., Liu, X., Swindeman, R., Brinkman, B.A.W., LeBlanc, M., Tsekenis, G., Friedman, N., Behringer, R., Denisov, D., Schall, P., Gu, X., Wright, W.J., Hufnagel, T., Jennings, A., Greer, J.R., Liaw, P.K., Becker, T., Dresen, G., Dahmen, K.A.: Universal quake statistics: from compressed nanocrystals to earthquakes. Sci. Rep. **5**, 16493 (2015)
29. Mäkinen, T., Miksic, A., Ovaska, M., Alava, M.J.: Avalanches in wood compression. Phys. Rev. Lett. **115**(5), 055501 (2015)
30. Deluca, A., Corral, Á.: Fitting and goodness-of-fit test of non-truncated and truncated power-law distributions. Acta Geophys. **61**(6), 1351–1394 (2013)
31. Clauset, A., Shalizi, C.R., Newman, M.E.J.: Power-law distributions in empirical data. SIAM Rev. **51**, 661–703 (2009)
32. Navas-Portella, V., Serra, I., González, Á., Vives, E., Corral, Á.: Universality of power-law exponents by means of maximum likelihood estimation. Phys. Rev. E **100**, 062106 (2019)
33. Pawitan, Y.: In All Likelihood: Statistical Modelling and Inference Using Likelihood. OUP Oxford, Oxford (2013)
34. Navas-Portella, V., Serra, I., Corral, Á., Vives, E.: Increasing power-law range in avalanche amplitude and energy distributions. Phys. Rev. E **97**, 022134 (2018)
35. Baró, J., Vives, E.: Analysis of power-law exponents by maximum-likelihood maps. Phys. Rev. E **85**(6), 066121 (2012)
36. Chou, T.-A., Dziewonski, A.M., Woodhouse, J.H.: Determination of earthquake source parameters from waveform data for studies of global and regional seismicity. J. Geophys. Res **86**, 2825–2852 (1981)
37. Nettles, M., Ekström, G., Dziewonski, A.M.: The Global CMT project 2004–2010: centroid-moment tensors for 13,017 earthquakes. Phys. Earth Planet. Inter. **200–201**, 1–9 (2012)
38. Yang, W., Hauksson, E., Shearer, P.M.: Computing a large refined catalog of focal mechanisms for Southern California (1981–2010): temporal stability of the style of faulting. Bull. Seismol. Soc. Am. **102**, 1179–1194 (2012)
39. Wessel, P., Luis, J., Uieda, L., Scharroo, R., Wobbe, F., Smith, W.H.F., Tian, D.: The generic mapping tools version 6. Geochem. Geophys. Geosyst. **6**(11), 5556–5564 (2019)
40. Hutton, K., Woessner, J., Hauksson, E.: Earthquake monitoring in Southern California for seventy-seven years (1932–2008). Bull. Seismol. Soc. Am. **100**(2), 423–446 (2010)
41. PCI-2—PCI-Based Two-Channel AE Board & System, by Physical Acoustics
42. Bender, B.: Maximum likelihood estimation of b values for magnitude grouped data. Bull. Seismol. Soc. Am. **73**(3), 831–851 (1983)
43. Pacheco, J.F.E., Scholz, C.H., Sykes, L.R.: Changes in frequency-size relationship from small to large earthquakes. Nature **355**, 71–73 (1992)
44. Yoder, M.R., Holliday, J.R., Turcotte, D.L., Rundle, J.B.: A geometric frequency-magnitude scaling transition: measuring $b = 1.5$ for large earthquakes. Tectonophysics **532–535**, 167–174 (2012)
45. Kamer, Y., Hiemer, S.: Data-driven spatial b value estimation with applications to California seismicity: to b or not to b. J. Geophys. Res.: Solid Earth **120**(7), 5191–5214 (2015)
46. Plenkers, K., Schorlemmer, D., Kwiatek, G., JAGUARS Research Group: On the probability of detecting picoseismicity. Bull. Seismol. Soc. Am. **101**(6), 2579–2591, 12 (2011)

47. Kwiatek, G., Plenkers, K., Nakatani, M., Yabe, Y., Dresen, G., JAGUARS-Group: Frequency-magnitude characteristics down to magnitude −4.4 for induced seismicity recorded at Mponeng gold mine. Bull. Seismol. Soc. Am. **100**(3), 1165–1173, 06 (2010)
48. Kagan, Y.: Universality of the seismic moment-frequency relation. Pure Appl. Geophys. **155**, 537–573 (1999)